NEUTRINOS:
A BIBLIOGRAPHY WITH INDEXES

NEUTRINOS:
A BIBLIOGRAPHY WITH INDEXES

ARNOLD S. LAVRO (EDITOR)

Nova Science Publishers, Inc.
New York

Senior Editors: Susan Boriotti and Donna Dennis
Coordinating Editor: Tatiana Shohov
Office Manager: Annette Hellinger
Graphics: Wanda Serrano
Editorial Production: Vladimir Klestov, Matthew Kozlowski and Maya Columbus
Circulation: Ave Maria Gonzalez, Vera Popovic, Luis Aviles, Raymond Davis,
        Melissa Diaz and Jeannie Pappas
Communications and Acquisitions: Serge P. Shohov
Marketing: Cathy DeGregory

Library of Congress Cataloging-in-Publication Data
Available upon Request

ISBN 1-59033-336-5.

CONTENTS

Preface   vii

Title Index   ix

Bibliography   1

Author Index   111

Subject Index   125

PREFACE

Neutrinos are one of the most abundant particles in the universe. Because they have very little interaction with matter, however, they are incredibly difficult to detect. Neutrinos are similar to the more familiar electron, with one crucial difference: neutrinos do not carry electric charge. Because neutrinos are electrically neutral, they are not affected by the electromagnetic forces which act on electrons. Three types of neutrinos are known. Each type or "flavor" of neutrino is related to a charged particle (which gives the corresponding neutrino its name). Hence, the "electron neutrino" is associated with the electron, and two other neutrinos are associated with heavier versions of the electron called the muon and the tau. This book presents citations from the literature for the last 3 years from the journal literature and the existent book literature. Access is provided by subject, author and title indexes.

TITLE INDEX

#

[187]Rhenium beta spectroscopy for neutrino mass determination: status report on the Genova experiment, 1

8 AgReO4 microcalorimeter array for the direct detection of the neutrino mass, 1

A

A 3-dimensional calculation of the atmospheric neutrino fluxes, 1

A Critical Examination of Sciama's Heavy Neutrino Hypothesis, 2

A dynamic solar core model: on the activity-related changes of the neutrino fluxes, 2

A frequentist analysis of solar neutrino data, 2

A General Parametrization for the Long-Range Part of Neutrinoless Double Beta Decay, 2

A Large Neutrino Detector Facility at the Spallation Neutron Source at Oak Ridge National Laboratory, 3

A Lead Astronomical Neutrino Detector: LAND, 3

A measurement of the neutrino-induced muon flux at the MACRO detector, 3

A mixed solar core, solar neutrinos and helioseismology, 3

A Model Independent Analysis of the Solar Neutrino Anomaly, 3

A Neutrino Component of Ultra High Energy Cosmic Rays?, 3

A Neutrino Factory, 3

A New Algorithm for Supernova Neutrino Transport and Some Applications, 3

A Novel Approach for Measuring the Beta-Neutrino Angular Correlation in Nuclear Beta Decay, 4

A novel approach in the detection of muon neutrino to tau neutrino oscillation from extragalactic neutrinos, 4

A phenomenological outlook on three-flavour atmospheric neutrino oscillations, 4

A Relic Neutrino Detector, 4

A Search for Lunar Cerenkov Emission from High-Energy Neutrinos, 4

A search for point sources of high-energy neutrinos with the AMANDA-B10 neutrino telescope, 4

A Signature of Solar Antineutrinos in Superkamiokande, 5

A Sterile Neutrino Need for Heavy-Element Nucleosynthesis, 5

A Study of 1- and 2-moment Closures for Supernova Neutrino Transport, 5

A Supernova Burst Observatory to Study μ and &tau, 5

Abundance and evolution of galaxy clusters in cosmological models with massive neutrino, 5

Accelerator Neutrino Oscillation Experiments, 6

Accelerator Test of the Dark Matter Neutrino Hypothesis, 6

Accurate GPS orientation of a long baseline for neutrino oscillation experiments at Fermilab, 6

Alternative Mechanisms for Neutrino Oscillations, 6

AMANDA and IceCube: Observatories for High-Energy Neutrino Astronomy, 6

An Updated Analysis on Atmospheric Neutrinos, 7

An upper bound to the high-energy neutrino flux from astrophysical sources, 7

An upper limit on the diffuse flux of high energy neutrinos obtained with the Baikal detector NT-96, 7

Analysis of the Neutrino Oscillation Based on the Seismic Solar Model, 7

Another Look at Just-So Solar Neutrino Oscillations, 8

Apparent Latitudinal Modulation of the Solar Neutrino Flux, 8

Are There Four or More Neutrinos?, 8

Are There Two Sterile Neutrinos Cooscillating with ν_e and ν_μ, 8

Aspects of massive neutrinos in astrophysics and cosmology, 9

Astronomie avec des neutrinos., 9

Astrophysical constraints on a possible neutrino ball at the Galactic Center, 9

Astrophysical Sources of High Energy Neutrinos, 9

Atmospheric and astrophysics neutrinos with MACRO, 10

Atmospheric Neutrino Observation in Super-Kamiokande - Evidence for ν_μ, 10

Atmospheric Neutrino Oscillations, 10

Atmospheric Neutrino Results from Soudan 2, 10

Atmospheric Neutrinos and Neutrino Oscillations, 10

Atmospheric Neutrinos from Charm, 10

Atmospheric neutrinos: phenomenological summary and outlook, 10

Automated Programming and Neutrino (Astro)Physics, 10

B

B and L Violation, Neutrino Mass and Oscillation, Proton Decay, 10

Background light in potential sites for the ANTARES undersea neutrino telescope, 10

Backreaction Effects of Dissipation in Neutrino Decoupling, 11

Baryogenesis through mixing of heavy Majorana neutrinos, 11

BOREXINO: a real-time detector for low-energy solar neutrinos, 11

Bounds on Neutrino Mixing Angles within the Context of SU (6)_L&otimes, 11

Bursts of Gravitational Waves Driven by Neutrino Oscillations, 11

C

Calibration of Sudbury Neutrino Observatory for the detection of boron-8 neutrinos, 11

Can ^{3}He redistribution solve the solar neutrino problem ?, 12

Can a supernova be located by its neutrinos?, 12

Can Parity Violation in Neutrino Transport Lead to Pulsar Kicks?, 12

Cherenkov emissions of Langmuir plasmons and light neutrinos by the flux of neutrinos, 12

Closure in flux-limited neutrino diffusion and two-moment transport, 12

Coherent Conversion of Neutrino Flavour by Collisions with Relic Neutrino Gas, 13

Coherent Neutrino Nucleus Scattering, 13

Collective neutrino-plasma interactions, 13

Collider Signatures of Sneutrino Cold Dark Matter, 13

Colliding neutron stars. Gravitational waves, neutrino emission, and gamma-ray bursts, 13

Comments on the neutrino oscillation scene, 14

Comments on the Physics Potential of Ultra-High Neutrino Interactions with Owl, 14

Comparative Analysis of GALLEX-GNO Neutrino Data and SXT X-Ray Data, 14

Comparative Analysis of GALLEX-GNO Neutrino Data, SOHO-MDI Helioseismology Data, and SXT X-ray Data, 15

Comparison of Neutrino Transfer Methods for 1d Supernova Simulations, 16

Conditions for shock revival by neutrino heating in core-collapse supernovae, 16

Confluence of cosmology, massive neutrinos, elementary particles, and gravitation, 17

Connection Between Relic Neutrinos and Cosmic Rays at >, 17

Constraining the window on sterile neutrinos as warm dark matter, 17

Constraints on R-Parity Breaking in GUT Constrained Mssm from the Neutrinoless Double Beta Decay, 17

Contribution of the Weak Interaction to Neutrino Thermodynamics in Astrophysical Plasmas, 17

Correlative Aspects of the Solar Electron Neutrino Flux and Solar Activity, 17

Cosmic microwave background anisotropy in the decaying neutrino cosmology, 17

Cosmic Neutrino Oscillations, 18

Cosmic Rays and Neutrinos from Gamma-Ray Bursts, 18

Cosmological Limits on the Neutrino Mass from the LyAlpha Forest, 18

Cosmological Neutrino Background Revisited, 19

Could intergalactic dust obscure a neutrino decay signature?, 19

CP, T violation in neutrino oscillations, 19

CP-Violation Effects in Long Baseline Neutrino Oscillation Experiments, 19

CP-Violation in Neutrino Oscillations: Theory, 19

Critical Solution in the Collapsing-Exploding Scenario of a Neutrino-Radiating Star, 19

Cryogenic Detectors for Radiochemical Solar Neutrino Experiments, 19

Current aspects of neutrino physics, 19

Current Status of Neutrino Oscillation Hunting - CHORUS and NOMAD, 20

Current status of the resonant spin-flavor precession solution to the solar neutrino problem, 20

D

Decaying Neutrinos and Large-Scale Structure Formation, 20

Decaying Neutrinos and the Flattering of the Galactic Halo, 20

Degenerate and Other Neutrino Mass Scenarios and Dark Matter, 20

Detecting Ultra High Energy Neutrinos by Upward Tau Airshowers and Gamma Flashes, 20

Detection of Very Small Neutrino Masses in Double-Beta Decay Using Laser Tagging, 20

Determining the Supernova Direction by Its Neutrinos, 20

Developing an Observatory for Multiflavor Neutrino Interactions from Supernovae (OMNIS), 21

Digital optical module and system design for km-scale neutrino detector in ice, 21

Direct measurements of neutrino mass, 21

Direct search for mass of neutrino and anomaly in the tritium beta-spectrum, 21

Do Statistically Significant Correlations Exist between the Homestake Solar Neutrino Data and Sunspots?, 22

Do We Understand the Origin and Impact of the Neutrino Flux in a Core Collapse Supernova?, 22

Dynamical QCD predictions for ultrahigh energy neutrino cross sections, 22

E

EeV Neutrinos, 23

Effect of Anisotropic Neutrino Radiation on Supernova Explosion Energy, 23

Effect of Coulomb collisions on time variations of the solar neutrino flux, 23

Effect of internal waves on the solar neutrino flux, 24

Effects of nuclear re-interactions in quasi-elastic neutrino-nucleus scattering, 24

Electromagnetic Neutrino Properties and Neutrino Oscillations in Electromagnetic Fields, 24

Electron Neutrino Mass Measurement by Supernova Neutrino Bursts and Implications on Hot Dark Matter, 24

Electron neutrino mass measurements by supernova neutrino bursts and implications for hot dark matter., 24

Electroweak baryon number violation and constraints on left-handed Majorana neutrino mass, 24

Electroweak Measurements and Neutrino Oscillations: The NuTeV and BooNE Experiments, 24

Element-Loaded Organic Scintillators for Neutron and Neutrino Physics, 25

Evidence against the Sciama Model of Radiative Decay of Massive Neutrinos, 25

Evidence for Neutrino Oscillation from Super-Kamiokande, 25

Evidence for Neutrino Oscillation Observed in Super-Kamiokande, 25

Evidence for Neutrino Oscillations from LSND: Past, Present and Future, 25

Evidence for Neutrino Oscillations in LSND, 26

Exotic mechanisms for the neutrino masses, 26

Experimental Program on the Future CERN-Gran Sasso Neutrino Beam-Line (NGS), 26

Experimental summary of the working group on neutrino oscillation, 26

Exploring the Neutrino-Driven Explosion Mechanism of Massive Stars, 26

Extragalactic neutrino background from very young pulsars surrounded by supernova envelopes, 26

Extreme-Energy Cosmic Rays: Puzzles, Models, and Maybe Neutrinos, 27

Extremely high energy neutrinos, neutrino hot dark matter, and the highest energy cosmic rays., 27

F

Field-theoretical treatment of neutrino oscillations, 27

Final Results of the Gallex Solar Neutrino Experiment, 27

First measurement of the flux of solar neutrinos from the sun at the Sudbury Neutrino Observatory, 27

First Neutrino Observations from the Sudbury Neutrino Observatory, 27

First results from the CHOOZ experiment on neutrino oscillations, 27

Flavor changing neutrino oscillations in LSND, 27

Flavor Physics: Kaons, Charm, Beauty, Taus, and Neutrinos, 27

Flavour models for neutrino physics, 28

Four-neutrino mixing and Big-Bang Nucleosynthesis, 28

Four-Neutrino Mixing, Oscillations and Big-Bang Nucleosynthesis, 28

Four-Neutrino Oscillations, 28

Four-neutrino spectrum from oscillation data, 28

Freezeout and Neutrinos in R-Process Nucleosynthesis, 28

Further Evidence for Neutrino Oscillations from Lsnd: the ν_{μ} --> &nu, 28

Future Dark Matter and Neutrino Physics Experiments at the UK Boulby Mine, 29

Future Detection of Supernova Neutrino Burst and Explosion Mechanism, 29

Future High Energy Neutrino Program at CERN, 29

Future Neutrino Astrophysics Projects at the UK Boulby Mine, 29

Future Prospects for Accelerator-Based Neutrino Mass Searches, 29

Future solar neutrino projects: Borexino, 29

G

Galactic Gamma Halo by Heavy Neutrino Annihilations?, 29

Galactic rotational velocities and dark neutrino matter, 29

GALLEX Solar Neutrino Results, 30

Gamma-ray and neutrino flares produced by protons accelerated on an accretion disc surface in active galactic nuclei, 30

Gamma-Ray Bursts via the Neutrino Emission from Heated Neutron Stars, 30

Gamma-Ray Bursts: Afterglow, High-Energy Cosmic Rays, and Neutrinos, 30

Gamma-rays and neutrinos from very young supernova remnants, 30

General Relativistic Augmentation of Neutrino Pair Annihilation Energy Deposition near Neutron Stars, 31

General Relativistic Effects on Neutrino-driven Winds from Young, Hot Neutron Stars and r-Process Nucleosynthesis, 31

General Relativistic Simulations of Stellar Core Collapse and Postbounce Evolution with Boltzmann Neutrino Transport, 31

Generation of neutrino masses and mixings in gauge theories, 31

GENIUS and the Genius TF: A New Observatory for WIMP Dark Matter and Neutrinoless Double Beta Decay, 31

Going Beyond the Salam-Weinberg Standard Model with Decaying Neutrinos, 31

Grand Observatories and Multiple-OWL for High-Energy Neutrino Astrophysics, 31

Gravitational Phase Transition of Heavy Neutrino Matter, 32

GRB990705: no evidence for neutrino signal., 32

Green's Functions for Neutrino Mixing, 32

H

Heavy neutrino ball as a possible solution to the "blackness problem", 32

Helioseismic Constraints on Solar Structure and the Solar Neutrino Problem, 32

Helioseismic constraints on the solar neutrino fluxes, 32

Helioseismology and solar neutrinos, 32

Helioseismology and Solar Neutrinos, 32

Helioseismology and the solar neutrino problem, 32

Hidden source of high-energy neutrinos in collapsing galactic nucleus, 33

High Energy Cosmic Neutrino Astronomy: The ANTARES Project, 33

High energy cosmic-rays and neutrinos from cosmological gamma-ray burst fireballs., 33

High Energy Neutrino Astronomy and the ANTARES Project, 33

High Energy Neutrino Astronomy with Antares, 33

High energy neutrino astronomy, 33, 34

High Energy Neutrino Astronomy: PASCOS 99, 33

High Energy Neutrino Astronomy: Towards Kilometer-Scale Detectors, 33, 34

High-energy astrophysics neutrinos at MACRO, 34

High-Energy Gamma and Neutrino Astronomy, 34

High-Energy Neutrino Astronomy: Towards Kilometer-Scale Detectors, 34

High-Energy Neutrinos from Photomeson Processes in Blazars, 34

High-energy physics: Neutrinos reveal split personalities, 34

Histogram Analysis of GALLEX, GNO and SAGE Neutrino Data: Further Evidence for Variability of the Solar Neutrino Flux, 34

Histogram Analysis of GALLEX, GNO, and SAGE Neutrino Data: Further Evidence for Variability of the Solar Neutrino Flux, 34

How Uncertain are Solar Neutrino Predictions?, 35

Hydrodynamical Study of Neutrino-Driven Wind as an r-Process Site, 35

I

ICECUBE: The Future of Neutrino Astronomy, 35

II. Neutrino Masses and Neutrino Oscillations, 35

Implications of Neutrino Flux Bounds for Future Neutrino Observations, 35

Implications of Recent Neutrino Oscillation Data, 36

Implications of the New Neutrino Oscillation Results, 36

Implications of the nu_mu-->, 36

Inclusive Neutrino and Antineutrino Reactions in Iron and an Interesting Relation, 36

Influence of Collective Plasma Processes on the Theoretical Flux of Solar Neutrinos, 36

In-Ice Radio Detection of GZK Neutrinos, 36

Invariant Box-Parameterization of Neutrino Oscillations, 36

Inverse neutrinoless double beta decay and &Delta, 36

IR limits, pregalactic stars, neutrino decay and quantum gravity, 36

Is neutrino decay really ruled out as a solution to the atmospheric neutrino problem from Super-Kamiokande data?, 37

Is There Time Variation of the Homestake Neutrino Flux?, 37

K

KamLAND Experiment-Exploring Low Energy Neutrino Physics, 37

Karmen ν_{e} --> &nu, 37

Korea Makes a Bid to Catch Neutrinos From the Cosmos, 37

L

Large neutrino flavour mixings and lepton mass matrices, 37

Lepton Asymmetry Generation from Netrino Oscillations in the Early Universe, 37

Limits on Neutrino Masses in SO(10)GUT'S, 37

Long Baseline Neutrino Oscillation Experiments, 37

Looking for Neutralinos From Neutrinos, 37

Low-metal core solar models with helioseismic constraints and the solar neutrino problem, 37

LSND Neutrino Oscillation Results and Implications, 38

M

MACRO and the Atmospheric Neutrino Problem, 38

Many-Valued Relationship Between Solar Flare Activity and Neutrino Flux in GALLEX, 38

Mass signature of supernova nu_{mu} and nu_{tau} neutrinos in SuperKamiokande, 38

Mass signature of supernova nu_{mu} and nu_{tau} neutrinos in the Sudbury Neutrino Observatory, 38

Mass splitting of three seesaw neutrinos, 39

Massive Neutrinos as Dark Matter ?? an Overview of Experiments, 39

Massive Neutrinos in a Coset-Space Family Unification, 39

Massive neutrinos in physics and astrophysics, 39

Massive Neutrinos, 39

Massive sterile neutrinos as warm dark matter, 39

Matter-affected neutrino oscillations in ordinary and mirror stars and their implications for gamma-ray bursts, 39

Measurement of the Solar Electron Neutrino Flux with the Homestake Chlorine Detector, 40

Measurements of ^{241}Pu ß-SPECTRUM in Search for Admixture of Massive Neutrinos, 40

MINOS: a long baseline neutrino oscillation experiment at NUMI, 40

Model Independent Analysis of the Solar Neutrino Data, 40

Modification of Neutrino Reaction Rates in Hot Dense Matter, 40

MSW Time Variations of the Solar Neutrino flux, 40

Mu- and Tau-Neutrino Spectra Formation in Supernovae, 41

Multi-dimensional Simulations of Core Collapse Supernovae employing Ray-by-Ray Neutrino Transport, 41

Multi-GEV Neutrinos from Internal Dissipation in Gamma-Ray Burst Fireballs, 41

Muon Measurements in the Atmosphere in the Context of the Atmospheric Neutrino Anomaly, 42

Muon Neutrinos with the MACRO Detector at L.N.G.S., 42

N

Nearly degenerate neutrino masses and nearly decoupled neutrino oscillations, 42

Neutral-Current Detection via ^{3}He (n.p)^{3}H in the Sudbury Neutrino Observatory, 42

Neutrino (Mu) -->, 42

Neutrino absorption tomography of the Earth's interior using isotropic ultra-high energy flux, 42

Neutrino Afterglow from Gamma-Ray Bursts: $\sim 10^{18}$ EV, 42

Neutrino Afterglows and Progenitors of Gamma-Ray Bursts, 42

Neutrino Annihilation between Binary Neutron Stars, 43

Neutrino Astronomy and Indirect Search for WIMPs, 43

Neutrino Astronomy and the Amanda South Pole Telescope, 43

Neutrino Astronomy in Experiment and Theory, 43

Neutrino Astronomy with the MACRO Detector, 43

Neutrino Astronomy: First Light-Using The South Pole IceCap, 44

Neutrino Astronomy: The Case for Antares, 44

Neutrino astrophysics and cosmology, 44

Neutrino Astrophysics at the Cross Roads, 45

Neutrino Astrophysics with the MACRO Detector, 45

Neutrino Breakout Bursts in Stellar Collapse, 45

Neutrino Burst from Supernovae and Neutrino Oscillation, 45

Neutrino Conversion and Neutrino Astrophysics, 45

Neutrino conversions in cosmological gamma-ray burst fireballs, 45

Neutrino Conversions in Solar Random Magnetic Fields, 45

Neutrino Cooling of Neutron Stars: Medium Effects, 45

Neutrino Cosmology, 46

Neutrino dark matter search at accelerators, 46

Neutrino Dark Matter, 46

Neutrino decay and the thermochemical equilibrium of the interstellar medium, 46

Neutrino deficit challenges conservation laws, 46

Neutrino Degeneracy and Chemically Inhomogeneous Universe, 46

Neutrino Degeneracy in Big-Bang Nucleosynthesis and Cosmic Rays: Evolution of the Light Elements, 46

Neutrino Detectives: Digging Deep at the Sudbury Neutrino Observatory, 46

Neutrino Electron Plasma Instability, 46, 47

Neutrino electron scattering and electroweak gauge structure: probing the masses of a new Z boson, 47

Neutrino emission due to Cooper pairing of nucleons in cooling neutron stars, 47

Neutrino Emission due to Cooper Pairing of Nucleons in Neutron Stars, 47

Neutrino emission due to proton pairing in neutron stars, 48

Neutrino emission from neutron stars, 48

Neutrino Event Rates from Gamma-Ray Bursts, 48

Neutrino Factory Detector and Long Baseline Oscillations, 48

Neutrino Flavor Mixing and Oscillations in Field Theory, 48

Neutrino flavor transformation in supernovae and the early universe, 48

Neutrino Fluence after r-Process Freezeout and Abundances of TE Isotopes in Presolar Diamonds, 48

Neutrino flux from observable Gamma Ray Bursts, 49

Neutrino Heating in an Inhomogeneous Big Bang Nucleosynthesis Model, 49

Neutrino Induced Waves in Degenerate Electron Plasmas, 49

Neutrino induced waves in degenerate electron plasmas: a mechanism in supernovae or gamma ray bursts?, 49

Neutrino Kinetics in Dense Astrophysical Plasmas, 49

Neutrino Magnetic Moment and Supernovae, 50

Neutrino magnetic moments and atmospheric neutrinos, 50

Neutrino Mass and Dark Matter, 50

Neutrino mass and its implications for the zero mode and vacuum structures of the standard model and its extensions, 50

Neutrino mass and oscillations, 50

Neutrino Mass from Tritium &beta, 50

Neutrino mass varying with time, 51

Neutrino Mass, 50, 51

Neutrino Masses and Leptogenesis from R Parity Violation, 51

Neutrino Masses and Mixing from Supersymmetric Inflation, 51

Neutrino Masses and Mixings: Big Bang and Supernova Nucleosynthesis and Neutrino Dark Matter, 51

Neutrino mixing and oscillations in 1999 and beyond, 51

Neutrino Mixing from Neutrino Oscillation Data, 51

Neutrino mixing schemes, 51

Neutrino nucleosynthesis (modern status), 51

Neutrino opacities at high density and the protoneutron star evolution, 51

Neutrino Oscillation at LSND, 51

Neutrino oscillation constraints on neutrinoless double-beta decay, 51

Neutrino Oscillation Effects in Indirect Detection of Dark Matter, 51

Neutrino oscillation experiments at nuclear reactors, 51

Neutrino Oscillation Results from Karmen, 52

Neutrino Oscillation Search in CHORUS and NOMAD, 52

Neutrino Oscillation Searches at CHORUS, 52

Neutrino oscillations and blazars, 52

Neutrino Oscillations and Cosmology, 52

Neutrino Oscillations and Elementarity in Leptonic Matter, 52

Neutrino Oscillations and Gamma-Ray Bursts, 52

Neutrino oscillations and the solar neutrino problem, 53

Neutrino Oscillations in Magnetized Media and Implications for the Pulsar Velocity Puzzle, 53

Neutrino oscillations in the early universe: how can large lepton asymmetry be generated?, 53

Neutrino Oscillations, 52, 53

Neutrino Oscillations: a phenomenological overview, 53

Neutrino oscillations: a source of Goldstone fields and consequences for supernovae, 53

Neutrino oscillations: accelerator experiments, 53

Neutrino Pair Annihilation above a Kerr Black Hole with the Accretion Disk, 53

Neutrino Pair Annihilation in the Gravitation of Gamma-Ray Burst Sources, 53

Neutrino Physics and Astrophysics, 54

Neutrino Physics and the Primordial Elemental Abundances, 54

Neutrino Physics with MACRO Detector, 54

Neutrino Physics with Thermal Detectors, 54

Neutrino Physics, 54

Neutrino Physics, Science of the New Millennium, 54

Neutrino Physics: Status and Prospect, 54

Neutrino Process Contributions to LiBeB Nucleosynthesis, 54

Neutrino Production by the Moon, 55

Neutrino propagation through dense matter, 55

Neutrino propagation through fluctuating media, 55

Neutrino Puzzles and Their Implications for the Nature of New Physics, 55

Neutrino Reactions in Nuclei and Neutrino Backgrounds, 55

Neutrino signals from WIMP annihilation, 55

Neutrino Spectroscopy with Karmen, 55

Neutrino spin flip in system of magnetic barriers, 55

Neutrino telescopes in Antarctica, 55

Neutrino Transport and Large-Scale Convection in Core-Collapse Supernovae, 55

Neutrino Transport in Type II Supernovae Boltzmann Solver vs. Monte Carlo Method, 56

Neutrino transport in type II supernovae: Boltzmann solver vs. Monte Carlo method, 56

Neutrino-driven Jets and Rapid-Process Nucleosynthesis, 57

Neutrino-driven supernovae: Boltzmann neutrino transport and the explosion mechanism, 57

Neutrino-Electron Scattering as a Probe of the Electroweak Gauge Structure, 57

Neutrino-induced Fission and r-Process Nucleosynthesis, 57

Neutrino-Induced Synthesis of ^{7}LI in He-Shell, 58

Neutrinoless Double Beta Decay Potential in a Large Mixing Angle World, 58

Neutrinoless double beta decay with Xe-136 in BOREXINO and the BOREXINO Counting Test Facility, 58

Neutrino-pair bremsstrahlung by electrons in neutron star crusts, 58

Neutrino-pair emission in a strong magnetic field, 58

Neutrinos after Takayama, 59

Neutrinos and Core Collapse Supernovae, 59

Neutrinos and Extra Dimensions, 59

Neutrinos and Muons, 59

Neutrinos and Nuclear Responses in Nuclear
    DOUBLE-ß and INVERSE-ß Processes, 59

Neutrinos and Supermassive Stars: Prospects for
    Neutrino Emission and Detection, 59

Neutrinos and supernova theory., 59

Neutrinos from active galactic nuclei as a
    diagnostic tool, 59

Neutrinos from active galactic nuclei, 59

Neutrinos from AGN, 60

Neutrinos from Blazars and the Differences
    between FRI and FRII Radio Galaxies, 60

Neutrinos from Cosmic Ray Interactions and
    Relativistic Astrophysical Sources, 60

Neutrinos from Early-Phase, Pulsar-driven
    Supernovae, 60

Neutrinos from Gamma-ray Bursts, 61

Neutrinos from Protoneutron Stars: Probing Hot
    and Dense Matter, 61

Neutrinos from the Sun, 61

Neutrinos in Cosmology, 61

Neutrinos in Physics and Astrophysics, 61

Neutrinos in physics, astrophysics, and
    cosmology, 61

Neutrinos in the Formation of Heavy Elements,
    62

Neutrinos, 59, 60, 61, 62

Neutrinos, cosmology and astrophysics, 62

Neutrinos, Gamma Rays and EUV, 62

New Directions for New Dimensions: From
    Strings to Neutrinos to Axions to, 62

New era in neutrino physics, 62

New Evidence for Neutrino Degeneracy in the
    Early Universe, 62

New Limits on a Diffuse Flux of >, 62

New Physics at Large Neutrino Telescopes, 62

New Results for Relevant Neutrino Emission in
    Superfluid Neutron Star Matter, 62

New results from the Mainz neutrino mass
    experiment, 62

New Soudan 2 Results on Neutrino Oscillations
    and Cosmic Rays, 62

Nonequilibrium Cosmic Neutrions and
    Nucleosynthesis, 63

Non-equilibrium Massive Tau-Neutrinos in the
    Early Universe, 63

Nonequilibrium Neutrinos in Cosmology, 63

Nonlinear Neutrino Interactions with the Stellar
    Envelope of a Supernova, 63

Nuclear Physics Issues in Neutrino Oscillation
    Experiments, 63

Nuclear Physics Limits to the ^{8}B Solar
    Neutrino Flux, 63

O

Observation of atmospheric muon neutrinos with
    AMANDA, 63

Observation of atmospheric neutrino events with
    the AMANDA experiment, 64

Observation of high-energy neutrinos using
    Cerenkov detectors embedded deep in
    Antarctic ice, 64

Observation of UHE Neutrino Interactions from
    Outer Space, 64

Observing the birth of supermassive black holes
    with the planned ICECUBE neutrino detector,
    64

On Steady State Neutrino-heated Ultrarelativistic
    Winds from Compact Objects, 65

On the detection of neutrino oscillations with
    Planck surveyor, 65

On the detection of ultra high energy neutrinos
    with the Auger observatory, 65

On the Energy of Neutrinos from Gamma-Ray
    Bursts, 65

On the Formation of Dark Matter Balls
    Composed of Degenerate, Self-gravitating
    Neutrinos, 66

On the Mass of the Neutrino, 66

On the Neutrino Flux from Gamma-Ray Bursts,
    66

On the possibility of radar echo detection of
    ultra-high energy cosmic ray- and neutrino-
    induced extensive air showers, 66

On the relation of extragalactic cosmic ray and
    neutrino fluxes, 66

On the resonant spin flavor precession of the
    neutrino in the sun, 67

On the Solar-Cycle Modulation of the Homestake
    Solar Neutrino Capture Rate and the Shuffle
    Test, 67

Optical Confirmation of a Neutrino-Detected
    Supernova, 67

Optical Emission Lines from Warm Interstellar
    Clouds: A Decisive Test of the Decaying
    Neutrino Theory, 67

Optimization of the Data Processing Algorithm
    of Searching for Neutrino-Gamma-Gravity
    Correlation, 68

Oscillating Neutrinos from the Galactic Center,
    68

Overview of Neutrino Oscillation Physics, 68

P

Parametric Resonance of Neutrino Oscillations in Electromagnetic Wave, 68

Parity Violation in Neutrino Transport and the Origin of Pulsar Kicks, 68

Particle physics, astrophysics and cosmology with forbidden neutrinos, 69

Particles in the Bulk: A Higher-Dimensional Approach to Neutrino and Axion Phenomenology, 69

Pauli's ghost: the conception and discovery of neutrinos, 69

Perturbative quantum chromodynamics predictions for very high-energy atmospheric neutrinos and muons, 69

Phenomenological Analysis of CP-Violation in Neutrino Oscillation Experiments, 69

Phenomenology of atmospheric neutrinos, 69

Plasma Acceleration of Photons (and Neutrinos), 69

Plasma instabilities driven by intense neutrino winds and anomalous heating in supernovae, 69

Ponderomotive Force of an Arbitrary Neutrino Distribution in a Magnetized Plasma, 70

Possible Relation between Relic Neutrinos and the Highest Energy Cosmic Rays, 70

Possible test for the suggestion that air showers with E>, 70

Possible Universal Neutrino Interaction, 70

Post-GZK Air Showers, FCNC, Strongly Interacting Neutrinos and Duality, 70

Predictions of the Solar Neutrino Fluxes and the Solar Gravity Mode Frequencies from the Solar Sound Speed Profile, 70

Present Status of the Oto Cosmo Observatory and Neutrino Physics by Means of ELEGANT Detectors, 71

Primary Cosmic Ray Flux and Neutrino Oscillations, 71

Primary Cosmic-Ray Spectrum and the Intensity of Atmospheric Neutrinos, 71

Primordial Nucleosynthesis and Neutrino Cosmology, 71

Primordial Nucleosynthesis with Neutrino Degeneracy and Gravitational Constant Variation, 71

Primordial Nucleosynthesis with Varying Gravitational Constant and Neutrino Degeneracy, 71

Primordially produced helium-4 in the presence of neutrino oscillations, 71

Probing Early Universe Physics with Cosmic, Gamma-Ray, and Neutrino Astrophysics, 71

Probing neutrino properties with the cosmic microwave background, 71

Probing the desert with ultra-energetic neutrinos from the sun and the earth, 72

Probing the interior of the Sun by measuring the solar neutrino flux, 72

Probing the Nature of Neutrinos at Pion Factories, 72

Probing Unstable Massive Neutrinos with Current CMB Observations, 72

Production and Detection of Black Holes Using a Neutrino Array, 73

Production of Light p-Process Isotopes in Neutrino-Irradiated Alpha-Rich Freezeouts, 73

Production of neutrons, neutrinos and gamma-rays by a very fast pulsar in the Galactic Centre region, 73

Progress Toward a Km-Scale Neutrino Detector in the Deep Ocean, 74

Progress towards OMNIS, the Observatory for Multi-flavor NeutrInos from Supernovae, 74

Propagation of Cosmic Rays and Neutrinos Through Space, 74

Prospects for Present and Future Reactor Neutrino Experiments, 74

Prospects for Radio Detection of Extremely High Energy Cosmic Rays and Neutrinos in the Moon, 74

Prospects of Hydroacoustic Detection of Ultra-High and Extremely High Energy Cosmic Neutrinos, 74

Prototype detector for ultrahigh energy neutrino detection, 74

Pulsar Acceleration by Asymmetric Emission of Sterile Neutrinos, 74

Pulsar Kicks from Asymmetric Neutrino Transport, 75

Pulsar Kicks from Neutrino Oscillations, 75

Pulsating pre-white dwarfs as laboratories for neutrino astrophysics, 75

R

Radiation of Angular Momentum by Neutrinos from Merged Binary Neutron Stars, 75

Real time supernova neutrino burst detection with MACRO, 75

Recent Development on Collective Neutrino Interactions, 76

Recent Results from Neutrino Oscillation Experiments at CERN, 76

Recent Results in Neutrino Masses, 76

Recent Super-Kamiokande Result on Atmospheric Neutrino and K2K Long Baseline Neutrino Oscillation Experiment, 76

Recent Super-Kamiokande Result on Solar Neutrinos, 76

Reexamination of Standard Solar Model to the Solar Neutrino Problems, 76

Registration of atmospheric neutrinos with the BAIKAL Neutrino Telescope NT-96, 76

Relativistic Effects on Neutrino Pair Annihilation above a Kerr Black Hole with the Accretion Disk, 77

Relic Neutrinos and Z-Resonance Mechanism for Highest Energy Cosmic Rays, 77

Relic Neutrinos, Monopoles, and Cosmic Rays above $\sim 10^{20}$ eV, 77

Resonance spin flavour precession and solar neutrinos, 77

Resonant active-sterile neutrino conversion and r-process nucleosynthesis in neutrino-heated supernova ejecta, 78

Results from Super-Kamiokande on Atmospheric Neutrinos, 78

Results from the NOMAD Neutrino Oscillation Experiment, 78

Results on 300 Days' Atmospheric Neutrino Data from Super-Kamiokande, 78

Results on neutrino oscillations from Super-Kamiokande, 78

Role of group and phase velocity in high-energy neutrino observatories, 78

Rotational and Related Periodicities in the Homestake and GALLEX Neutrino Data, 78

Rotational Signature and Possible R-Mode Signature in the GALLEX Solar Neutrino Data, 79

S

Science and technology of Borexino: a real-time detector for low energy solar neutrinos, 79

Scintillation Crystal Detector for Low Energy Neutrino and Astroparticle Physics, 80

Search for a possible space-time correlation between high energy neutrinos and gamma-ray bursts, 80

Search for EHE Neutrinos with the the HiRes Fly's Eye (Stage 1) detector., 80

Search for neutral heavy leptons in a high-energy neutrino beam, 80

Search for neutrino decay during the 1999 solar eclipse, 81

Search for Neutrino Mass and Oscillations with a Powerful Supernova Detector Array, 81

Search for supernova neutrino bursts with the AMANDA detector, 81

Search for the Antineutrino Rest Mass in the Tritium Beta Decay, 82

Search for WIMPs at neutrino telescopes, 82

Searches for neutrino mass in the NuTeV experiment, 82

Searches for neutrino oscillations: LSND, 82

Searching for Low Energy Electron Anti-Neutrinos from the Sun, 82

Searching for the MSW Solar Neutrino Oscillation Enhancement Using the Earth, 82

Seasonal Variations in Solar High-Energy Neutrino Flux and Their Probable Source, 82

Secondary decays in atmospheric charm contributions to the flux of muons and muon neutrinos, 82

Seismic constraints on neutrino oscillation parameters, 83

Seismic solar models and the neutrino problem, 83

Semi-analytic approximations for production of atmospheric muons and neutrinos, 83

Sensitivity of an underwater acoustic array to ultra-high energy neutrinos, 83

Sensitivity of Borexino to Seasonal Variations of the Solar Neutrino Flux, 83

Shadows of Relic Neutrino Masses and Spectra on Highest Energy GZK Cosmic Rays, 83

Simulation of Neutrino Transport by Large-Scale Convective Instability in a Proto-Neutron Star, 83

Simulations of Core Collapse Supernovae in One and Two Dimensions Using Multigroup Neutrino Transport, 83

Sneutrino physics with lepton number violation, 84

SNEWS: A Neutrino Early Warning System for Galactic SN II, 84

SNO Observables from the Magnetic Moment Solution to the Solar Neutrino Problem, 84

Solar and Atmospheric Neutrino Oscillations - Super-Kamiokande Results, 84

Solar and Atmospheric Neutrino Results from Super-Kamiokande, 84

Solar and Atmospheric Neutrinos, 84

Solar and Supernova Constraints of Cosmologically Interesting Neutrinos, 84

Solar Magnetic Field from Solar Neutrino Experiments, 84

Solar Models Based on Helioseismology and the Solar Neutrino Problem, 84

Solar Models with Helioseismic Constraints and the Solar Neutrino Problem, 85

Solar models: constraints from helioseismology and neutrino production, 85

Solar Models: Current Epoch and Time Dependences, Neutrinos, and Helioseismological Properties, 85

Solar neutrino detection: recent results and future perspectives, 85

Solar Neutrino Emission Deduced from a Seismic Model, 85

Solar neutrino experiment: current status and perspectives, 86

Solar neutrino in relation to solar wind particles, 86

Solar Neutrino Observation in Super-Kamiokande, 86

Solar Neutrino Results from Super-Kamiokande, 86

Solar Neutrino Spectroscopy with Borexino and Recent Results from the Ctf Experiment, 86

Solar Neutrinos as a Possible Diagnostic of the Solar Magnetic Field, 87

Solar Neutrinos as Highlight of Astroparticle Physics, 87

Solar Neutrinos, 86, 87

Solar neutrinos, atmospheric neutrinos and proton decays in Super-Kamiokande and the Kam-LAND project, 87

Solar Neutrinos: An Overview, 87

Solar Neutrinos: Where We are and What is Next?, 87

Solar Neutrinos: Where We Are, 87

Solar Structure After Neutrinos and Helioseismology (CD-ROM Directory: contribs/gough), 87

Solar-Neutrino Problem Solved, 87

Some Particle Physics Aspects of Neutrinoless Double Beta Decay, 87

Some Remarks on the Neutrino Oscillation Phase in a Gravitational Field, 87

Some Topics in Cosmology and Neutrino Physics, 87

Spectrum of solar neutrinos above 6.5 MeV, 88

Spherical collapse of supermassive stars: Neutrino emission and gamma-ray bursts, 88

Spherically Symmetric Core Collapse Supernova Simulations With Boltzmann Neutrino Transport, 89

Spin-Flavour Conversions of Neutrinos in Collapsing Stars, 90

Standard Physics Solution to the Solar Neutrino Problem?, 90

Stars as galactic neutrino sources, 90

Status and Perspectives of Neutrino Oscillation Searches, 90

Status and Perspectives of the Mainz Neutrino Mass Experiment, 90

Status of Neutrino Oscillation Searches, 90

Status of the Milano neutrino mass experiment with thermal detectors, 90

Status of the Neutrino Telescope AMANDA: Monopoles and WIMPS, 90

Status of Trimaximal Neutrino Mixing, 90

Sterile Neutrinos and CMB, 90

Sterile Neutrinos in Big Bang Nucleosynthesis, 90

Sterile Neutrinos: Phenomenology and Theory, 90

Studies of neutrino oscillations at accelerators, 90

Studies of neutrino oscillations at reactors, 91

Studies of the Sudbury Neutrino Observatory detector and sonoluminescence using a sonoluminescent source, 91

Studies Towards a Detector for Extragalactic SuperNova Neutrinos, 91

Study of Salt Neutrino Detector, 91

Sudbury Neutrino Observatory energy calibration using gamma-ray sources, 91

Super-Kamiokande 0. 07 eV Neutrinos in Cosmology: Hot Dark Matter and the Highest Energy Cosmic Rays, 92

Supernova Bounds on Neutrino Properties: A Mini-Review, 92

Supernova II Neutrino Bursts and Neutrino Massive Mixing, 92

Supernova neutrino detection in Borexino, 92

Supernova Neutrino Opacity from Nucleon-Nucleon Bremsstrahlung and Related Processes, 92

Supernova Neutrino Oscillation in the Presence of Random Magnetic Field, 93

Supernova neutrinos., 93

Supernovae, Neutrinos, and Amateur Astronomers, 93

T

Tau neutrinos in the Auger Observatory: a new window to UHECR sources, 93

Technique for Direct eV-Scale Measurements of the Mu and Tau Neutrino Masses Using Supernova Neutrinos, 93

TeV Neutrinos and GeV Photons from Shock Breakout in Supernovae, 93

Textures for Neutrino Mass Matrices in Gauge Theories, 94

Textures of Neutrino Mass Matrix with Large Flavor Mixing, 94

The ^{51}Cr and ^{90}Sr sources in BOREXINO as tools for neutrino magnetic moment searches, 94

The AMANDA Neutrino Experiment and its Expansion to 1 Kilometer Dimension, 94

The AMANDA Neutrino Telescope, 94

The AMANDA Neutrino Telescope: Design, Construction, and Performance, 94

The AMANDA Neutrino Telescope: Science Prospects and Performance at First Light, 94

The AMANDA Neutrino Telescope: Status and Latest Results, 94

The AMANDA South Pole Neutrino Telescope: First Light, 95

The Antares Demonstrator. Towards a High Energy Undersea Neutrino Telescope, 95

The ANTARES Neutrino Telescope: Status and Prospects, 95

The atmospheric neutrino anomaly: muon neutrino disappearance, 95

The Baikal Deep Under Water Neutrino Experiment: Results, Status, Future, 95

The BAIKAL Neutrino Project: Status Report, 95

The C/CO ratio in B335 and the decaying dark matter neutrino hypothesis, 95

The Case for a Neutrino-Degenerate Universe, 95

The Case of the Missing Neutrinos: Results from SuperKamiokande, 96

The CERN Neutrino Oscillation Experiments, 96

The CERN-LNGS neutrino programme, 96

The CUORE project: a large observatory for neutrinoless double beta decay and other rare events, 96

The Design and Status of the Sudbury Neutrino Observatory, 96

The Development and Calibration of An(127)I Solar Neutrino Detector, 96

The east-west effect for atmospheric neutrinos, 97

The electron scattering reaction in the Sudbury Neutrino Observatory, 97

The Evolution of Neutrino Astronomy, 98

The Extent and Cause of the Pre-White Dwarf Instability Strip, with Applications to Neutrino Astrophysics, 98

The Fermilab Neutrino Oscillation Facility, 98

The fluxes of sub-cutoff particles detected by AMS, the cosmic ray albedo and atmospheric neutrinos, 98

The Galaxy Distribution and the Hubble Law in a Neutrino Dominated Universe, 99

The geometry of atmospheric neutrino production, 99

The Highest Energy Cosmics Rays, Photons, and Neutrinos, 99

The history of solar neutrino problem, 99

The Interplay between Proto--Neutron Star Convection and Neutrino Transport in Core-Collapse Supernovae, 99

The jet-disk symbiosis model for gamma ray bursts: cosmic ray and neutrino background contribution, 100

The long baseline neutrino oscillation experiment from KEK to Super-Kamiokande - k2k, 100

The nature of massive neutrinos, 100

The neutrino mass from tritium &beta, 100

The Neutrino Telescope ANTARES, 100

The neutrino-induced neutron source in helium shell and r-process nucleosynthesis, 100

The Neutrino-Oscillation Experiment at the Chooz Nuclear Power Plant, 101

The Nuclear Physics of Solar and Supernova Neutrino Detection, 101

The Oscillation Length Resonance in the Transitions of Solar and Atmospheric Neutrinos Crossing the Earth Core, 101

The Palo Verde Reactor Neutrino Experiment. a Test for Long Baseline Neutrino Oscillations, 101

The Palo Vere Neutrino Oscillation Experiment, 101

The Pattern of Neutrino Masses and How to Determine It, 101

The Predicted Signature of Neutrino Emission in Observations of Pulsating Pre-White Dwarf Stars, 101

The Proposed ORLaND Neutrino Facility, 101

The Role of Neutrinos, Strings, Gravity, and Variable Cosmological Constant in Elementary Particle Physics, 102

The r-Process in Neutrino-driven Winds from Nascent, "Compact" Neutron Stars of Core-Collapse Supernovae, 102

The r-Process Nucleosynthesis in Neutrino-/Magnetocentrifugally-Driven Winds, 102

The SIREN Solar Neutrino Experiment, 102

The solar neutrino problem after three hundred days of data at SuperKamiokande, 102

The Solar Neutrino Problem: Mixing of Neutrinos and Mixing in the Sun, 103

The Solar Neutrino Puzzle, 103

The solar neutrino puzzle: the way ahead, 103

The standard solar model and the neutrino problem: present status and future perspectives, 103

The Status of Neutrino Mass, 103

The status of neutrino physics, 103

The Sudbury Neutrino Observatory, 103

The sun as a high energy neutrino source, 104

The Sun, a laboratory for neutrino- and astrophysics, 104

The Two Gravitating Massive Neutrino Pairs, 104

The Weight of Neutrinos and Related Questions, 104

Theoretical Possibilities and Observational Constraints for Radiatively Decaying Neutrinos with Mass Near 30 EV, 104

Theory of neutrino masses and mixings, 104

Thermodynamical instability of self-gravitating heavy neutrino matter, 104

Thermodynamics of the Neutrino Gas in a Plasma, 104

Three Flavor Neutrino Oscillation Analysis of the Superkamiokande Atmospheric Neutrino Data, 104

Three-Neutrino Vacuum Oscillation Solutions to the Solar and Atmospheric Anomalies, 105

Time Variation of the Solar Neutrino Flux and Correlations with Solar Phenomena, 105

Top Five Reasons for Rejecting the Weyl-Majorana Massless Neutrino Confusion Theorem, 105

Towards the resolution of the solar neutrino problem, 105

Two-Neutrino Double Beta Decay: a Study of Different Approximation Schemes, 105

Type-II supernovae and neutrino magnetic moments, 105

U

Ultra High Energy Neutrinos by Tau Airshowers, 105

Ultra High Energy Neutrinos from Gamma Ray Bursts, 106

Ultrahigh Energy Neutrinos as Probe for Weak-scale String Theories?, 106

Ultrahigh energy neutrinos from gamma ray bursts., 106

Uncertainty of the solar neutrino energy spectrum, 106

Underwater-Ice Neutrino Telescopes, 106

Unitarity Constraints on Neutrino Mass and Mixings, 106

Updated Parameters for the Decaying Neutrino Theory and E, 106

Upper Bound on the Neutrino Magnetic Moment from Collisions Induced by Landau Damping in Supernovae, 106

V

Vacuum energy and configuration energy for a massive neutrino propagating in a neutron star, 107

Vacuum oscillations and excess of high energy solar neutrino events observed in Superkamiokande, 107

Vacuum oscillations and variations of solar neutrino rates in SuperKamiokande and Borexino, 107

Variability of the Solar Neutrino Flux, 107

Variations of the core luminosity and solar neutrino fluxes, 108

Very High Energy Gamma-Rays and Neutrinos from AGN, 108

VHE and UHE neutrinos from blazar flares, 108

VHE and UHE neutrinos from GRB - energy and flux limits, 108

W

Weak Interactions and Neutrinos, 109

Weighing neutrinos: weak lensing approach, 109

What can long-baseline neutrino oscillation experiments tell us about neutrino dark matter?, 109

With Neutrino Masses Revealed, Proton Decay is the Missing Link, 109

BIBLIOGRAPHY

"Signature" Neutrinos from Photon Sources at
High Redshift
Postma, M.
AIP Conf. Proc. 579: Radio Detection of
High Energy Particles, 2001/1/1, 53
Abstract - Not Available

^{187}Rhenium beta spectroscopy for neutrino
mass determination: status report on the
Genova experiment
Vitale, S. et al.
Weak Interactions and Neutrinos,
2000/1/1, 165
Abstract - Not Available

8 AgReO4 microcalorimeter array for the direct
detection of the neutrino mass
Alessandrello, Angelo; Brofferio, Chiara;
Cremonesi, Oliviero; Fiorini, Ettore;
Giuliani, Andrea; Nucciotti, Angelo;
Pavan, Maura; Pessina, Gianluigi;
Previtali, Ezio; Sisti, Monica; Zanotti,
Luigi; Margesin, Benno; Pigmatel,
Giorgio; Zen, Mario; Beeman, Jeffrey W.;
Haller, Eugene E.
Proc. SPIE Vol. 4140, p. 402-406, X-Ray
and Gamma-Ray Instrumentation for
Astronomy XI, Kathryn A. Flanagan;
Oswald H. Siegmund; Eds., 4140,
2000/12/1, 402-406
Abstract - We are presenting our recent
developments to measure the electron
antineutrino mass by studying the ^{187}Re
(beta) - spectrum end-point with high
resolution thermal detectors. We will
discuss the preliminary results of an array
of 8 bolometers made up of AgReO$_4$
absorbers (2.309 mg of total mass
corresponding to a total ^{187}Re active mass
of about 0.905 mg of with an expected
(beta) total rate of about 1.3 Hz). Their
risetime of 0.7 - 1.2 ms together with their
energy resolution, ranging between 21 eV
and 26 eV at 1.5 KeV, should allow to set a
limit of about 10 - 12 eV after one year of
real time measurements.

A 3-dimensional calculation of the atmospheric
neutrino fluxes
Battistoni, G.
Astroparticle Physics, 12, 2000/1/1, 315-
333
Abstract - Not Available

A Comparison of Boltzmann and Multigroup
Flux-limited Diffusion Neutrino Transport
during the Postbounce Shock Reheating
Phase in Core-Collapse Supernovae
Messer, O. E. B.; Mezzacappa, A.; Bruenn,
S. W.; Guidry, M. W.
Astrophysical Journal, 507, 1998/11/1,
353-360
Abstract - We compare Newtonian three-
flavor multigroup Boltzmann (MGBT) and
(Bruenn's) multigroup flux-limited
diffusion (MGFLD) neutrino transport in
postbounce core-collapse supernova
environments. We focus our study on
quantities central to the postbounce
neutrino heating mechanism for reviving
the stalled shock. Stationary-state three-
flavor neutrino distributions are developed
in thermally and hydrodynamically frozen
time slices obtained from core collapse and
bounce simulations that implement
Lagrangian hydrodynamics and MGFLD

neutrino transport. We obtain distributions for time slices at 106 and 233 ms after core bounce for the core of a 15 M_&sun; progenitor, and at 156 ms after core bounce for a 25 M_&sun; progenitor. For both transport methods, the electron neutrino and antineutrino luminosities, rms energies, and mean inverse flux factors, all of which enter the neutrino heating rates, are computed as functions of radius and compared. The net neutrino heating rates are also computed as functions of radius and compared. Notably, we find significant differences in neutrino luminosities and mean inverse flux factors between the two transport methods for both precollapse models and for all three time slices. In each case, the luminosities for each transport method begin to diverge above the neutrinospheres, where the MGBT luminosities become larger than their MGFLD counterparts, finally settling to a constant difference maintained to the edge of the core. We discuss the ramifications that these new results have for the supernova mechanism.

A Critical Examination of Sciama's Heavy Neutrino Hypothesis
Seahra, S. S.; Overduin, J. M.; Duley, W. W.; Wesson, P. S.
AIP Conf. Proc. 493: General Relativity and Relativistic Astrophysics, 1999/1/1, 219
Abstract - Not Available

A dynamic solar core model: on the activity-related changes of the neutrino fluxes
Grandpierre, Attila
Astronomy and Astrophysics, 348, 1999/8/1, 993-999
Abstract – The author points out that the energy sources of the Sun may actually involve runaway nuclear reactions as well, developed by the fundamental thermonuclear instability present in stellar energy producing regions. In this paper the author considers the conjectures of the derived model for the solar neutrino fluxes in case of a solar core allowed to vary in relation to the surface activity cycle. The observed neutrino flux data suggest a solar core possibly varying in time. In the dynamic solar model the solar core involves a " quasi-static" energy source produced by the quiet core with a lower than standard temperature which may vary in time. Moreover, the solar core involves another, dynamic energy source, which also changes in time. The sum of the two different energy sources may produce quasi-constant flux in the SuperKamiokande because it is sensitive to neutral currents, anti-neutrinos (and axions), therefore it observes the sum of the neutrino flux of two sources which together produce the solar luminosity. A dynamic solar core model is developed to calculate the contributions of the runaway source to the individual neutrino detectors. The results of the dynamic solar model are consistent with the present helioseismic measurements and can be checked with future helioseismic measurements as well. In the Appendix, the physical parameters of the bubbles are derived, including their temperatures, energy contents, sizes, velocities, lifetimes, and the possible lengths of path they can travel in the Sun.

A frequentist analysis of solar neutrino data
Garzelli, M. V.; Giunti, C.
Astroparticle Physics, 17, 2002/5/1, 205-220
Abstract - We estimate with Monte Carlo the goodness of fit and the confidence level of the standard allowed regions for the neutrino oscillation parameters obtained from the fit of solar neutrino data. The Monte Carlo estimates are significantly smaller than the corresponding standard values. Using Neyman's method, we also calculate exact allowed regions with correct frequentist coverage assuming the standard least-squares estimator of the oscillation parameters. Our results show that the standard allowed region around the global minimum of the least-squares function is a reasonable approximation of the exact one, whereas the size of the other regions is underestimated in the standard method.

A General Parametrization for the Long-Range Part of Neutrinoless Double Beta Decay
Päs, H.; Hirsch, M.; Kovalenko, S. G.;

Klapdor-Kleingrothaus, H. V.
Particle and Nuclear Physics, 1998/1/1,
283
Abstract - Not Available

A Large Neutrino Detector Facility at the
Spallation Neutron Source at Oak Ridge
National Laboratory
Efremenko, Yu. V.
Electroweak Physics, 2000/1/1, 343
Abstract - Not Available

A Lead Astronomical Neutrino Detector:
LAND
Hargrove, C. K.; Dubeau, J.
Particles and the Universe, 1998/1/1, 347
Abstract - Not Available

A measurement of the neutrino-induced muon
flux at the MACRO detector
Hanson, Kael Dylan Ph.D.
Thesis, 2000/9/1, 8
Abstract - Data from the MACRO detector
was utilized to perform a measurement of
the neutrino-induced muon flux in the high
energy range <E $_{\nu}$> ~ 100 GeV.
This measurement comprises a large
sample (403.6 events after all fidelity cuts
and background subtraction) from an
exposure time of 4 years. This is contrasted
with an expected event rate of 540 upward
muons from a detailed Monte Carlo
simulation of MACRO using as input the
atmospheric neutrino flux of the Bartol
group. This gives a ratio of 0.76 +/-
0.04(stat.) +/- 0.16(syst.) measured versus
expected. Recent results from other
experiments worldwide, most notably the
Super-Kamioka collaboration in Japan,
have exposed similar discrepancies in the
measured versus expected fluxes of
underground muons. A consistent
explanation of these effects is that the
neutrino flux undergoes flavor oscillations
between source and detector. For simple
two-flavor mixing, a mixing angle of sin
$^{2}2\theta_s$ ~ 1 and a mass-squared
difference in the range 0.001 <
Δm^2 < 0.01 eV^2 best fits the
observed muon neutrino flux.

A mixed solar core, solar neutrinos and
helioseismology

degl'Innocenti, S.; Ricci, B.
Astroparticle Physics, 8, 1998/4/1, 293-296
Abstract - We consider a wide class of
solar models with mixed core. Most of
these models can be excluded as the sound
speed profile that they predict is in sharp
disagreement with helioseismic constraints.
All the remaining models predict ^8B
and/or ^7Be neutrino fluxes of at least as
large as those of SSMs. In conclusion,
helioseismology shows that a mixed solar
core cannot account for the neutrino deficit
implied by the current solar neutrino
experiments.

A Model Independent Analysis of the Solar
Neutrino Anomaly
Heeger, K. M.; Robertson, R. G. H.
Particle and Nuclear Physics, 1998/1/1,
135
Abstract - Not Available

A Neutrino Component of Ultra High Energy
Cosmic Rays?
Domokos, G.; Kovesi-Domokos, S.
The Role of Neutrinos, Strings, Gravity,
and Variable Cosmological Constant in
Elementary Particle Physics, 2001/1/1, 103
Abstract - Not Available

A Neutrino Factory
Edgecock, R.
Identification of Dark Matter, 2001/1/1,
557
Abstract - Not Available

A New Algorithm for Supernova Neutrino
Transport and Some Applications
Burrows, Adam; Young, Timothy; Pinto,
Philip; Eastman, Ron; Thompson, Todd A.
Astrophysical Journal, 539, 2000/8/1, 865-
887
Abstract - We have developed an implicit,
multigroup, time-dependent, spherical
neutrino transport code based on the
Feautrier variables, the tangent-ray
method, and accelerated &b.Lambda;
iteration. The code achieves high angular
resolution, is good to O(v/c), is equivalent
to a Boltzmann solver (without
gravitational redshifts), and solves the
transport equation at all optical depths with
precision. In this paper, we present our

formulation of the relevant numerics and microphysics and explore protoneutron star atmospheres for snapshot postbounce models. Our major focus is on spectra, neutrino-matter heating rates, Eddington factors, angular distributions, and phase-space occupancies. In addition, we investigate the influence on neutrino spectra and heating of final-state electron blocking, stimulated absorption, velocity terms in the transport equation, neutrino-nucleon scattering asymmetry, and weak magnetism and recoil effects. Furthermore, we compare the emergent spectra and heating rates obtained using full transport with those obtained using representative flux-limited transport formulations to gauge their accuracy and viability. Finally, we derive useful formulae for the neutrino source strength due to nucleon-nucleon bremsstrahlung and determine bremsstrahlung's influence on the emergent ν_μ and ν_τ neutrino spectra. These studies are in preparation for new calculations of spherically symmetric core-collapse supernovae, proto-neutron star winds, and neutrino signals.

A Novel Approach for Measuring the Beta-Neutrino Angular Correlation in Nuclear Beta Decay
Beck, M.
Particles, Strings and Cosmology, 2000/1/1, 544
Abstract - Not Available

A novel approach in the detection of muon neutrino to tau neutrino oscillation from extragalactic neutrinos
Iyer, Sharada Ramlingam Ph.D.
Thesis, 2001/12/1, 2
Abstract - A novel approach is proposed for studying the ν_μ --> ν_τ oscillation and detection of extragalactic neutrinos. Active Galactic Nuclei (AGN), Gamma Ray Bursters (GRB) and Topological Defects are believed to be sources of ultrahigh energy ν_μ and ν_τ. These astrophysical sources provide a long baseline of 100Mpc, or more, for possible detection of ν_μ --> ν_τ oscillation with mixing parameter Δm^2 down to 10^{-17} eV^2,

many orders of magnitude below the current accelerator experiments. The propagation characteristics of upward going muon and tau neutrinos is studied to show that high energy tau neutrinos cascade down in energy as they propagate through the Earth, producing an enhancement of the incoming tau neutrino flux in the low energy region. By contrast, high energy muon neutrinos get attenuated as they traverse the Earth. An understanding of tau energy loss at very high energies could help with the interpretation of long tracks produced by charged particles in large underground detectors.

A phenomenological outlook on three-flavour atmospheric neutrino oscillations
Fogli, G. L. et al.
Weak Interactions and Neutrinos, 2000/1/1, 360
Abstract - Not Available

A Relic Neutrino Detector
Hagmann, C.
AIP Conf. Proc. 478: COSMO-98, 1999/1/1, 460
Abstract - Not Available

A Search for Lunar Cerenkov Emission from High-Energy Neutrinos
Hankins, T. H.; Ekers, R. D.; O'Sullivan, J. D.
AIP Conf. Proc. 579: Radio Detection of High Energy Particles, 2001/1/1, 168
Abstract - Not Available

A search for point sources of high-energy neutrinos with the AMANDA-B10 neutrino telescope
Young, Scott Matthew Ph.D.
Thesis, 2001/9/1, 2
Abstract - This dissertation describes a search for astronomical point sources of high energy neutrinos using the AMANDA- B10 detector. Good sensitivity is achieved over most of the northern hemisphere by tailoring the analysis to hard neutrino spectra (E^{-2}) and relaxing signal purity requirements. This strategy, thus far unique in AMANDA, produces large effective area and the lowest flux

limits. The data collected between April to November of 1997 (total of 130 days of livetime) has been analyzed. No point source candidates were identified. For sources with E^{-2} spectra, the detector achieves 10,000 m^2 in average muon effective area between declinations of 35 to 90 degrees. Flux limits for declinations larger than +45 degrees are competitive with the best limits in the northern sky. The impact of systematic errors on flux limits were determined by modifying detector related parameters such as OM average sensitivity and angular dependent sensitivity, ice models, etc. These studies confirm our predicted sensitivity to within 40%.

A Signature of Solar Antineutrinos in Superkamiokande
Fiorentini, G.; Moretti, M.; Villante, F. L.
Particle and Nuclear Physics, 1998/1/1, 149
Abstract - Not Available

A Sterile Neutrino Need for Heavy-Element Nucleosynthesis
Caldwell, David O.
Particles, Strings and Cosmology, 2000/1/1, 334
Abstract - Not Available

A Study of 1- and 2-moment Closures for Supernova Neutrino Transport
Young, T. R.; Burrows, A.; Pinto, P.
American Astronomical Society Meeting, 195, 1999/12/1
Abstract - We present results of 1-moment (flux-limited diffusion) and 2-moment closures for static neutrino atmospheres in the cores of massive stars. The code used in all calculations is an implicit, multi-group, spherical neutrino transport algorithm. Given these various closures, we calculate emergent spectra, luminosities, and net heating rates. We also compare the results using the various closures with those using a more precise technique to gauge the accuracy of such approximations.

A Supernova Burst Observatory to Study μ and τ Neutrinos

Boyd, R. N.
Eighteenth Texas Symposium on Relativistic Astrophysics, 1998/1/1, 726
Abstract - Not Available

A three generation oscillation analysis of the Super-Kamiokande atmospheric neutrino data beyond one mass scale dominance approximation
Choubey, Sandhya; Goswami, Srubabati; Kar, Kamales
Astroparticle Physics, 17, 2002/4/1, 51-73
Abstract - In this paper we do a three-generation oscillation analysis of the latest (1144 days) Super-Kamiokande (SK) atmospheric neutrino data going beyond the one mass scale dominance (OMSD) approximation. We fix Δ$_{12}$=Δ$_{13}$ (Δ$_{;LSND}$) in the range eV^2 as allowed by the results from LSND and other accelerator and reactor experiments on neutrino oscillation and keep Δ$_{23}$ (Δ$_{;ATM}$) and the three mixing angles as free parameters. We incorporate the matter effects, indicate some new allowed regions with small Δ$_{23}$ (/<$10^{-4}eV^2$) and $sin^2$2θ$_{23}$ close to 0 and discuss the differences with the two-generation and OMSD pictures. Implications for future long baseline experiments are discussed.

Abundance and evolution of galaxy clusters in cosmological models with massive neutrino
Arhipova, N. A.; Kahniashvili, T.; Lukash, V. N.
Astronomy and Astrophysics, 386, 2002/5/1, 775-783
Abstract - The time evolution of the number density of galaxy clusters and their mass and temperature functions are used to constrain cosmological parameters in the spatially flat dark matter models containing hot particles (massive neutrino) as well as cold and baryonic matter. We test the modified MDM (Lambda =0) models with cosmic gravitational waves and show that they neither pass the cluster evolution test nor reproduce the observed height of the first acoustic peak in Delta T/T spectrum, and therefore should be ruled out. The models with a non-zero cosmological

constant are in better agreement with observations. We estimate the free cosmological parameters in Lambda MDM with a negligible abundance of gravitational waves, and find that within the parameter ranges hin (0.6, 0.7), nin (0.9, 1.1), f_{nu} equiv $Omega_{nu}$ /$Omega_m$ in (0, 0.2), (i) the value of $Omega_{Lambda}$ is strongly affected by a small fraction of hot dark matter: ;0.45 <$Omega_{Lambda}$ <0.7 (1sigma CL), and (ii) the redshift evolution of galaxy clusters alone reveals the following explicit relation between $Omega_{Lambda}$ and f_{nu} : \Omega_ \Lambda +0.5f_ \nu =0.65\pm 0.1. This degeneracy is also expected in LSS tests (with a smaller error). The present accuracy of observational data allows to bound the fraction of hot matter, f_nu in (0, 0.2); the number of massive neutrino species remains undelimited, N_{nu} = 1, 2, 3.

Accelerator Neutrino Oscillation Experiments
Bazarko, Andrew O.
Particles, Strings and Cosmology,
2000/1/1, 361
Abstract - Not Available

Accelerator Test of the Dark Matter Neutrino
Hypothesis
Minakata, Hisakazu
Recent Developments in Theoretical and
Experimental General Relativity,
Gravitation, and Relativistic Field
Theories, 1999/1/1, 1441
Abstract - Not Available

Accurate GPS orientation of a long baseline for
neutrino oscillation experiments at
Fermilab
Soler, Tomás; Foote, Richard H.; Hoyle,
Dixon; Bocean, Virgil
Geophysical Research Letters, 27,
2000/12/1, 3921
Abstract - Not Available

Alternative Mechanisms for Neutrino
Oscillations
Halprin, A.
AIP Conf. Proc. 444: Particle Physics and
Cosmology, First Tropical Workshop,
1998/1/1, 40

Abstract - Not Available

AMANDA and IceCube: Observatories for
High-Energy Neutrino Astronomy
Price, P. B.
American Astronomical Society Meeting,
195, 1999/12/1
Abstract - High energy (E > 100 GeV) neutrino astrophysics, still a virtually unexplored frontier, has great discovery potential for particle astronomy, particle physics, cosmology, and even gravitational physics. The Antarctic Muon and Neutrino Detector Array (AMANDA), with an effective area of 10,000 square meters for TeV neutrinos traveling along its vertical axis, is now recording 100 neutrino-induced muons per year, predominantly of atmospheric origin. IceCube (effective area = 1 square kilometer) at the South Pole will use extraterrestrial neutrinos to * probe the origin of cosmic ray nuclei, the nature of gamma ray bursts, and the physical processes in the central engines of active galaxies; * search for point sources of TeV neutrinos and for annihilation products of cold dark matter in the sun and earth; * monitor the sky for Galactic supernovae and other cataclysmic phenomena. The talk will focus on AMANDA, IceCube, and the astronomical questions they may answer.

An Investigation of Neutrino-driven
Convection and the Core Collapse
Supernova Mechanism Using Multigroup
Neutrino Transport
Mezzacappa, A.; Calder, A. C.; Bruenn, S.
W.; Blondin, J. M.; Guidry, M. W.;
Strayer, M. R.; Umar, A. S.
Astrophysical Journal, 495, 1998/3/1, 911
Abstract - We investigate neutrino-driven convection in core collapse supernovae and its ramifications for the explosion mechanism. We begin with a postbounce model that is optimistic in two important respects: (1) we begin with a 15 M&sun; precollapse model, which is representative of the class of stars with compact iron cores; (2) we implement Newtonian gravity. Our precollapse model is evolved through core collapse and bounce in one dimension using multigroup (neutrino energy-dependent) flux-limited diffusion

(MGFLD) neutrino transport and Newtonian Lagrangian hydrodynamics, providing realistic initial conditions for the postbounce convection and evolution. Our two-dimensional simulation begins at 12 ms after bounce and proceeds for 500 ms. We couple two-dimensional piecewise parabolic method (PPM) hydrodynamics to precalculated one-dimensional MGFLD neutrino transport. (The neutrino distributions used for matter heating and deleptonization in our two-dimensional run are obtained from an accompanying one-dimensional simulation. The accuracy of this approximation is assessed.)

An Updated Analysis on Atmospheric Neutrinos
Gonzalez-Garcia, M. C.; Nunokawa, H.; Peres, O.; Stanev, T.; Valle, J. W. F.
Particle and Nuclear Physics, 1998/1/1, 251
Abstract - Not Available

An upper bound to the high-energy neutrino flux from astrophysical sources
Waxman, E.; Bahcall, J. N.
Abstracts of the 19th Texas Symposium on Relativistic Astrophysics and Cosmology, held in Paris, France, Dec. 14-18, 1998. Eds.: J. Paul, T. Montmerle, and E. Aubourg (CEA Saclay)., 1998/12/1, 618
Abstract - An upper bound to the high-energy neutrino flux from astrophysical sources, imposed by cosmic-ray observations, is derived. The upper bound applies to sources which are optically thin to proton photo-meson interaction, and cannot be avoided by invoking source redshift evolution or inter-galactic magnetic fields. This upper limit is two orders of magnitude below the flux predicted in popular AGN jet models, but is consistent with our predictions from GRB models. Its implications to future neutrino detectors are discussed.

An upper limit on the diffuse flux of high energy neutrinos obtained with the Baikal detector NT-96
Balkanov, V. A.; Belolaptikov, I. A.; Bezrukov, L. B.; Budnev, N. M.; Chensky, A. G.; Danilchenko, I. A.; Dzhilkibaev, Z.- A. M.; Domogatsky, G. V.; Doroshenko, A. A.; Fialkovsky, S. V.; Gaponenko, O. N.; Kiss, D.; Klabukov, A. M.; Klimov, A. I.; Klimushin, S. I.; Koshechkin, A. P.; Kuznetzov, V. E.; Kulepov, V. F.; Kuzmichev, L. A.; Ljaudenskaite, J. J.; Lovzov, S. V.; Lubsandorzhiev, B. K.; Milenin, M. B.; Mirgazov, R. R.; Moseiko, N. I.; Netikov, V. A.; Osipova, E. A.; Panfilov, A. I.; Parfenov, Y. V.; Pavlov, A. A.; Pliskovsky, E. N.; Pokhil, P. G.; Popova, E. G.; Rozanov, M. I.; Rubzov, V. Y.; Sokalski, I. A.; Spiering, C.; Streicher, O.; Tarashansky, B. A.; Tothi, G.; Thon, T.; Vasiljev, R.; Wischnewski, R.; Yashin, I. V.
Astroparticle Physics, 14, 2000/9/1, 61-66
Abstract - We present the results of a search for high-energy neutrinos with the Baikal underwater Cherenkov detector NT-96. An upper limit on the diffuse flux of ν_e+ ν_μ+$\bar{\nu_\mu}$ of $E^2\Phi_\nu(E) < 1.4 \times 10^{-5} cm^{-2} s^{-1} sr^{-1} GeV$ within neutrino energy range $10^{4-107} GeV$ is obtained, assuming an E^{-2} behavior of the neutrino spectrum.

Analysis of the Neutrino Oscillation Based on the Seismic Solar Model
Takata, M.; Hosogai, M.; Shibahashi, H.
SOHO-9 Workshop "Helioseismic Diagnostics of Solar Convection and Activity", Stanford, California, July 12-15, 1999., 9, 1999/1/1, 84
Abstract - By adopting the seismic solar model constructed with a constraint of the sound speed profile and the density profile determined by helioseismology, we evaluate the neutrino oscillation hypothesis (the MSW effect) as a possible solution to the solar neutrino problem and determine the range of the parameters of this hypothesis so that they are not inconsistent with the solar neutrino flux measurement (Homestake, GALLEX, SAGE and Super-Kamiokande). The parameters are (i) the squared mass difference between the electron neutrino and the mu neutrino and (ii) the mixing angle between the mass eigenstate and the flavor eigenstate. We find that there are three allowed regions of the parameter space as in the case of the analysis based on the evolutionary solar

models. Each region is, however, slightly shifted as a result of a small difference between the seismic solar model and the evolutionary solar model. We calculate the neutrino energy spectrum for each of these three regions, and compare them with the result of the Super-Kamiokande experiment.

Another Look at Just-So Solar Neutrino Oscillations
Gelb, J. M.; Rosen, S. P.
American Astronomical Society Meeting, 193, 1998/12/1
Abstract - We take another look at "Just-So" solar neutrino oscillations and characterize them by the energy at which the distance-varying angle is pi /2. The SuperKamiokande spectrum can be fit with an energy in the range of 9-12 MeV. In this case, the pp neutrinos are reduced to half the standard solar model prediction and (7) Be neutrinos must make up a significant part of the gallium signal, in contrast to the small-angle MSW prediction. Furthermore, the (7) Be neutrinos could show a large seasonal variation, which might be detectable in more precise gallium data.

Apparent Latitudinal Modulation of the Solar Neutrino Flux
Sturrock, P. A.; Walther, G.; Wheatland, M. S.
Astrophysical Journal, 507, 1998/11/1, 978-983
Abstract - We examine the solar neutrino flux, as measured by the Homestake neutrino detector, to search for evidence of a dependence upon the solar latitude of the Earth-Sun line that varies from 7.25d south in mid-March to 7.25d north in mid-September. Although the flux does not obviously show any dependence on latitude, we do find evidence for a dependence of the variance of the flux upon latitude. When data from 108 runs of the Homestake experiment are divided into four quartiles, sorted according to latitude, we find that the northernmost quartile exhibits a larger variance than the other three. By applying the shuffle test, we estimate the probability that this could have occurred by chance to be in the range

1%-2%. For more detailed information, we examine a "reconstructed flux" formed from our recent maximum likelihood spectrum analysis. This procedure indicates that the variance is largest at about 6.5d north. We also find that the spectrum of the variance of the reconstructed flux has a notable peak at 1 cycle y^-1 tending to confirm a latitude dependence of the variance. We also examine the 12.88 cycle yr periodicity described in our recent paper and find that the amplitude of the periodicity is greater for the northernmost quartile than for the other quartiles. We suggest that these effects may be attributed to resonant spin-flavor precession of left-hand-helicity electron neutrinos in the magnetic field of the solar radiative zone.

Are There Four or More Neutrinos?
Whisnant, Kerry
Confluence of Cosmology, Massive Neutrinos, Elementary Particles, and Gravitation, 1999/1/1, 63
Abstract - Not Available

Are There Two Sterile Neutrinos Cooscillating with ν_e and ν_μ?
Królikowski, W.
LNP Vol. 539: Theoretical Physics Fin de Siècle, 2000/1/1, 251
Abstract - The existence of two sterile neutrinos ν_s and ν'_s (blind to all Standard-Model interactions) is shown to be implied by a model of fermion "texture" that we develop since some time. They may mix nearly maximally with two of three conventional neutrinos, say ν_e and ν_μ, thus leading to neutrino oscillations, say ν_e --> ν_s and ν_μ --> ν'_s, with nearly maximal amplitudes. Then, they can be responsible for the observed deficits of solar ν_e's and atmospheric ν_μ's, respectively, but by themselves do not help to explain the LSND results for ν_μ -->

ν_e oscillations. On the other hand, they are consistent with the CHOOZ negative result. At the moment, the experiment cannot decide, whether the deficit of atmospheric ν_μ's, confirmed by the recent findings, has to be related to the oscillations ν_μ --> ν_τ or ν_μ --> ν'_s. In the last Section of the paper, a new notion of "non-Abelian spin-1/2 fermions" is presented in the context of a composite option for fermion families.

Aspects of massive neutrinos in astrophysics and cosmology
Keranen, Petteri Ph.D.
Thesis, 1999/1/1, 20
Abstract - The objective of this thesis is to study neutrino properties in the framework of astrophysics and cosmology. The research is concentrated on phenomena connected with the active galactic nuclei (AGN) that are supposed to be the origin of highest energy cosmic rays and ultra-high energy neutrinos. The recent SuperKamiokande results on the atmospheric neutrino flux strongly indicate that neutrinos have a mass and that they mix with each other. This discovery has made the study of neutrino properties particularly topical. The observation of very high-energy neutrinos that are supposed to be produced in AGN or related objects would provide a new probe for testing oscillations, electromagnetic properties, decays and beyond-the-SM interaction forms of neutrinos. It is shown that one may obtain stricter constraints on the strength of exotic neutrino-neutrino interactions than the existing limits from laboratory experiments and other astrophysical observations. It is also shown that the observation of AGN neutrino flux may reveal important information about neutrino magnetic moments. The long travelling distance of AGN neutrinos allows for a sensitive test of neutrino instability. The decay of a heavy neutrino into a light neutrino and a scalar is studied, and it is shown that the appropriate

coupling strength may be tested in an accuracy several orders of magnitude better than in other phenomena. Also cosmology provides information of the basic neutrino properties. The thermalization of the wrong-helicity Dirac neutrinos is studied by paying attention in particular to the pole effects of the weak gauge bosons. By solving numerically the relevant Boltzmann equation the abundance of wrong-helicity neutrinos is evaluated.

Astronomie avec des neutrinos.
Moscoso, L.
Journal des Astronomes Francais, 56, 1998/1/1, 15
Abstract - Not Available

Astrophysical constraints on a possible neutrino ball at the Galactic Center
De Paolis, F.; Ingrosso, G.; Nucita, A. A.; Orlando, D.; Capozziello, S.; Iovane, G.
Astronomy and Astrophysics, 376, 2001/9/1, 853-860
Abstract - The nature of the massive object at the Galactic Center (Sgr A*) is still unclear even if various observational campaigns led many Authors to believe that our Galaxy hosts a super-massive black hole with mass $M = \sim 2.6 \times 10^6\ M_{sun}$. However, the black hole hypothesis, which theoretically implies a luminosity $= \sim 10^{41}$ erg s-1, runs into problems if one takes into account that the observed luminosity, from radio to gamma -ray wavelengths, is below 10^{37} erg s-1. In order to solve this blackness problem, alternative models have recently been proposed. In particular, it has been suggested that the Galactic Center hosts a ball made up of non-baryonic matter ({e.g.} massive neutrinos and anti-neutrinos) in which the degeneracy pressure of fermions balances their self-gravity. Requiring it to be consistent with all the available observations towards the Galactic Center allows us to put severe astrophysical constraints on the neutrino ball parameters.

Astrophysical Sources of High Energy Neutrinos
Protheroe, R. J.

Towards the Millennium in Astrophysics,
Problems and Prospects. International
School of Cosmic Ray Astrophysics 10th
Course, 1998/1/1, 149
Abstract - Not Available

Atmospheric and astrophysics neutrinos with
MACRO
Montaruli, T.; The MACRO Collaboration
Dark matter in Astrophysics and Particle
Physics, 1999/1/1, 790
Abstract - Not Available

Atmospheric Neutrino Observation in Super-
Kamiokande - Evidence for ν μ
Oscillations
Kajita, T.; The Super-KAMIOKANDE
Collaboration
New Era in Neutrino Physics , 1998/1/1,
107
Abstract - Not Available

Atmospheric Neutrino Oscillations
Casper, D. W.
AIP Conf. Proc. 478: COSMO-98,
1999/1/1, 399
Abstract - Not Available

Atmospheric Neutrino Results from Soudan 2
Litchfield, P. J.
Particle and Nuclear Physics, 1998/1/1,
205
Abstract - Not Available

Atmospheric Neutrinos and Neutrino
Oscillations
Stanev, Todor
New Vistas in Astrophysics, 2000/1/1, 65
Abstract - Not Available

Atmospheric Neutrinos from Charm
Pasquali, L.; Reno, M. H.; Sarcevic, I.
AIP Conf. Proc. 478: COSMO-98,
1999/1/1, 408
Abstract - Not Available

Atmospheric neutrinos: phenomenological
summary and outlook
Lipari, P.
Weak Interactions and Neutrinos,
2000/1/1, 370
Abstract - Not Available

Automated Programming and Neutrino
(Astro)Physics
Maris, Michele
ASP Conf. Ser. 172: Astronomical Data
Analysis Software and Systems VIII, 8,
1999/1/1, 50
Abstract - A project for the creation of a
code for accurate numerical predictions for
solar neutrino oscillations is presented. The
project was realized through an automated
program generator which scans the
simulation parameter space. The software
validation procedure and the production
guidelines are illustrated.

B and L Violation, Neutrino Mass and
Oscillation, Proton Decay
Lepton and Baryon Number Violation in
Particle Physics, neral Astrophysics and
Cosmology, 1999/1/1, 109
Abstract - Not Available

Background light in potential sites for the
ANTARES undersea neutrino telescope
Amram, P.; Anvar, S.; Aslanides, E.;
Aubert, J.-J.; Azoulay, R.; Basa, S.;
Benhammou, Y.; Bernard, F.; Bertin, V.;
Billault, M.; Blanc, P.-E.; Blanc, F.; Bland,
R. W.; Blondeau, F.; Bottu, N.; Boulesteix,
J.; Brooks, B.; Brunner, J.; Calzas, A.;
Carloganu, C.; Carmona, E.; Carr, J.;
Carton, P.-H.; Cartwright, S.; Cases, R.;
Cassol, F.; Compere, C.; Cooper, S.;
Coustillier, G.; de Botton, N.; Deck, P.;
Desages, F. E.; Destelle, J.-J.; Dispau, G.;
Drogou, J. F.; Drouhin, F.; Duval, P.-Y.;
Feinstein, F.; Festy, D.; Fopma, J.; Fuda,
J.-L.; Goret, P.; Gosset, L.; Gournay, J.-F.;
Hernández, J. J.; Herrouin, G.; Hubaut, F.;
Hubbard, J. R.; Huss, D.; Jaquet, M.;
Jelley, N.; Kajfasz, E.; Karolak, M.;
Kouchner, A.; Kudryavtsev, V.; Lachartre,
D.; Lafoux, H.; Lamare, P.; Languillat, J.-
C.; Laugier, D.; Laugier, J.-P.; Le Guen,
Y.; Le Provost, H.; Le Van Suu, A.;
Lemoine, L.; Liotard, P. L.; Loucatos, S.;
Magnier, P.; Macelin, M.; Martin, L.;
Massol, A.; Mazeau, B.; Mazure, A.;
Mazéas, F.; McMillan, J.; Millot, C.; Mols,
P.; Montanet, F.; Morel, J. P.; Moscoso, L.;
Navas, S.; Olivetto, C.; Palanque-
Delabrouille, N.; Pallares, A.; Payre, P.;
Perrin, P.; Pohl, A.; Poinsignon, J.;

Potheau, R.; Queinec, Y.; Racca, C.; Raymond, M.; Rolin, J. F.; Sacquin, Y.; Schuller, J.-P.; Schuster, W.; Spooner, N.; Stolarczyk, T.; Tabary, A.; Talby, M.; Tao, C.; Tayalati, Y.; Thompson, L. F.; Triay, R.; Tzvetanov, T.; Valdy, P.; Vernin, P.; Vigeolas, E.; Vignaud, D.; Vilanova, D.; Wark, D.; Zghiche, A.; Zúñiga, J.
Astroparticle Physics, 13, 2000/5/1, 127-136
Abstract - The ANTARES collaboration has performed a series of in-situ measurements to study the background light for a planned undersea neutrino telescope. Such background can be caused by ⁴⁰K decays or by biological activity. We report on measurements at two sites in the Mediterranean Sea at depths of 2400 m and 2700 m, respectively. Three photomultiplier tubes were used to measure single counting rates and coincidence rates for pairs of tubes at various distances. The background rate is seen to consist of three components: a constant rate due to ^{40}K decays, a continuum rate that varies on a time scale of several hours simultaneously over distances up to at least 40 m, and random bursts a few seconds long that are only correlated in time over distances of the order of a meter. A trigger requiring coincidences between nearby photomultiplier tubes should reduce the trigger rate for a neutrino telescope to a manageable level with only a small loss in efficiency.

Backreaction Effects of Dissipation in Neutrino Decoupling
Maartens, Roy; Triginer, Josep
General Relativity and Gravitation, 32, 2000/9/1, 1711-1725
Abstract - Not Available

Baryogenesis through mixing of heavy Majorana neutrinos
Pilaftsis, A.
Dark matter in Astrophysics and Particle Physics, 1999/1/1, 93
Abstract - Not Available

BOREXINO: a real-time detector for low-energy solar neutrinos
Giammarchi, M. G.; BOREXINO Collaboration
Lepton and Baryon Number Violation in Particle Physics, Astrophysics and Cosmology, 1999/1/1, 210
Abstract - Not Available

Bounds on Neutrino Mixing Angles within the Context of SU (6)$_L$ ⊗U(1)$_Y$ Model
Gaitán, R.; García, E.; Hernández-Galeana, A.; Rivera-Rebolledo, J. M.
AIP Conf. Proc. 531: Particles and Fields, 2000/1/1, 337
Abstract - Not Available

Bursts of Gravitational Waves Driven by Neutrino Oscillations
Mosquera Cuesta, Herman J.
Astrophysical Journal, 544, 2000/11/1, L61-L64
Abstract - Active-to-sterile neutrino oscillations and nonspherical distortion of the resonance surface may trigger asymmetric emission of sterile neutrinos during the core bounce of a supernova collapse. The huge binding energy released and the proto-neutron star rapid rotation may power bursts of gravitational waves by the time the neutrino flavor conversions ensue. These bursts would be detectable up to distances of ~2.2 Mpc. The relativistic requirement of an ellipsoidal axisymmetric core at maximum gravitational wave emission may be explained by the neutrinosphere geometry at the oscillations' onset. Thus, neutrino oscillations might naturally be the underlying mechanism producing asymmetric structures, neutron star kicks, and bipolar jet ejecta in supernovae.

Calibration of Sudbury Neutrino Observatory for the detection of boron-8 neutrinos
Ford, Richard James Ph.D.
Thesis, 1999/8/1, 5
Abstract - The Sudbury Neutrino Observatory (SNO) is a second generation water Cerenkov detector using 1000 tonnes of heavy water to study neutrino astrophysics. Using deuterium neutrino reactions, SNO will measure the flux and

energy spectrum of solar electron neutrinos, and will measure the flavour-blind flux of neutrinos.

Can ^{3}He redistribution solve the solar neutrino problem ?
Antia, H. M.; Chitre, S. M.
Bulletin of the Astronomical Society of India, 27, 1999/1/1, 69
Abstract - Not Available

Can a supernova be located by its neutrinos?
Beacom, J. F.; Vogel, P.
Physical Review D, 60, 1999/8/1, 3007
Abstract - A future core-collapse supernova in our Galaxy will be detected by several neutrino detectors around the world. The neutrinos escape from the supernova core over several seconds from the time of collapse, unlike the electromagnetic radiation, emitted from the envelope, which is delayed by a time of order hours. In addition, the electromagnetic radiation can be obscured by dust in the intervening interstellar space. The question therefore arises whether a supernova can be located by its neutrinos alone. The early warning of a supernova and its location might allow greatly improved astronomical observations. The theme of the present work is a careful and realistic assessment of this question, taking into account the statistical significance of the various neutrino signals.

Can Parity Violation in Neutrino Transport Lead to Pulsar Kicks?
Arras, Phil; Lai, Dong
Astrophysical Journal, 519, 1999/7/1, 745-749
Abstract - In magnetized proto-neutron stars, neutrino cross sections depend asymmetrically on the neutrino momenta because of parity violation. However, these asymmetric opacities do not induce any asymmetric flux in the bulk interior of the star where neutrinos are nearly in thermal equilibrium. Consequently, parity violation in neutrino absorption and scattering can only give rise to asymmetric neutrino flux above the neutrino-matter decoupling layer. We estimate that a dipole field of

B~10^15-10^16 G is necessary to produce kick velocity of order of a few hundred km s^-1 using this mechanism.

Cherenkov emissions of Langmuir plasmons and light neutrinos by the flux of neutrinos
Tsintsadze, Levan N.
Astrophysics and Space Science, 274, 2000/1/1, 719-723
Abstract - We consider the system of neutrino-antineutrino (\nu \bar{\nu}) - plasma taking into account their weak Fermi interaction, when the density of neutrinos ν and the plasma particles are rather high and their collective interaction plays a vital role. New fluid instabilities driven by strong neutrino flux in a plasma are observed. It is shown that a bunch of neutrinos, drifting with a constant velocity across a homogeneous plasma, can also induce emission of lower energy neutrinos due to scattering, i.e. the decay of a heavy neutrino ν_H into a heavy and a light neutrino ν_L (ν_H-->ν_Hν_L) in a plasma. Furthermore we find that the neutrino production in stars does not lead in general to energy losses from the neutron stars.

Closure in flux-limited neutrino diffusion and two-moment transport
Smit, J. M.; van den Horn, L. J.; Bludman, S. A.
Astronomy and Astrophysics, 356, 2000/4/1, 559-569
Abstract - Two-dimensional maximum entropy closure and various standard one-dimensional closures in flux-limited neutrino diffusion and two-moment transport are compared against direct numerical solutions of the neutrino Boltzmann equation. The approximate transport based on particular closures of the moment equations is rated by testing so-called weak equivalence of the first three moments of the neutrino radiation field. Additionally we consider strong equivalence of the maximum entropy angular model distribution. Our calculations are performed on two different matter backgrounds and involve several neutrino energies. As an alternative multiple energy test we look at the

behavior of spectral and energy-averaged Eddington factors. Among the closures considered, two-dimensional maximum entropy closure is found to overall approximate most closely the full transport solutions.

Coherent Conversion of Neutrino Flavour by Collisions with Relic Neutrino Gas
Batkin, I. S.; Sundaresan, M. K.
AIP Conf. Proc. 488: High Energy Physics at the Millennium: MRST '99,, 1999/1/1, 110
Abstract - Not Available

Coherent Neutrino Nucleus Scattering
Oberauer, L.; Jochum, J.
Proceedings of the Sixth SFB-375 Ringberg Workshop Astroteilchenphysik, 2000/2/1, 6
Abstract - Not Available

Collective neutrino-plasma interactions
Tsytovich, V. N.; Bingham, R.; Dawson, J. M.; Bethe, H. A.
Astroparticle Physics, 8, 1998/4/1, 297-307
Abstract - Nonlinear processes describing the interaction of neutrinos with collective plasma oscillations and the excitation of plasma turbulence by a large neutrino flux is discussed. The excitation considered is the inverse processes of neutrino emission by plasma waves first considered by Tsytovich (V.N. Tsytovich, Soviet Fiz. Dokl. 9 (1965) 1114). The process is similar to a beam plasma instability considered as inverse Landau damping in which the usual electromagnetic interactions are important. In the neutrino beam relaxation the weak interaction can play a similar role. We emphasize here the possibility of another process namely the interaction of an intense neutrino flux with a strongly turbulent plasma. The turbulence can also be assumed to be due to the shock produced at the early stages of a type II supernova (SN) explosion. The scattering of the neutrinos in the turbulent plasma is shown to be sufficient for transferring momentum and energy from the neutrino flux to the plasma causing the shock to continue moving outward and eventually creating the blow-off of the mantle of the star producing type II SN.

Collider Signatures of Sneutrino Cold Dark Matter
Kolb, S. et al.
Dark Matter in Astro- and Particle Physics, 2001/1/1, 276
Abstract - Not Available

Colliding neutron stars. Gravitational waves, neutrino emission, and gamma-ray bursts
Ruffert, M.; Janka, H.-Th.
Astronomy and Astrophysics, 338, 1998/10/1, 535-555
Abstract - Three-dimensional hydrodynamical simulations are presented for the direct head-on or off-center collision of two neutron stars, employing a basically Newtonian PPM code but including the emission of gravitational waves and their back-reaction on the hydrodynamical flow. A physical nuclear equation of state is used that allows us to follow the thermodynamical evolution of the stellar matter and to compute the emission of neutrinos. Predicted gravitational wave signals, luminosities and waveforms, are presented. The models are evaluated for their implications for gamma-ray burst scenarios. We find an extremely luminous outburst of neutrinos with a peak luminosity of more than 4* 10(54) erg/s for several milliseconds. This leads to an efficiency of about 1% for the annihilation of neutrinos with antineutrinos, corresponding to an average energy deposition rate of more than 10(52) erg/s and a total energy of about 10(50) erg deposited in electron-positron pairs around the collision site within 10 ms. Although these numbers seem very favorable for gamma-ray burst scenarios, the pollution of the e(+/-) pair-plasma cloud with nearly 10(-1} M_{sun) of dynamically ejected baryons is 5 orders of magnitude too large. Therefore the formation of a relativistically expanding fireball that leads to a gamma-ray burst powered by neutrino emission from colliding neutron stars is definitely ruled out.

Comment on "Neutrino oscillations in the early universe: how can large lepton asymmetry

be generated?" [Astropart. Phys. 14 (2000) 79-90]
Di Bari, P.; Foot, R.; Volkas, R. R.; Wong, Y. Y. Y.
Astroparticle Physics, 15, 2001/8/1, 391-412
Abstract - We comment on the recent paper by A.D. Dolgov, S.H. Hansen, S. Pastor and D.V. Semikoz (DHPS) [Astropart. Phys. 14 (2000) 79] on the generation of neutrino asymmetries from active-sterile neutrino oscillations. We demonstrate that the approximate asymmetry evolution equation obtained therein is an expansion, up to a minor discrepancy, of the well-established static approximation equation, valid only when the supposedly new higher order correction term is small. In the regime where this so-called "back-reaction" term is large and artificially terminates the asymmetry growth, their evolution equation ceases to be a faithful approximation to the quantum kinetic equations simply because pure Mikheyev-Smirnov-Wolfenstein (MSW) transitions have been neglected. At low temperatures the MSW effect is the dominant asymmetry amplifier. Neither the static nor the DHPS approach contains this important physics. Therefore we conclude that the DHPS results have sufficient veracity at the onset of explosive asymmetry generation, but are invalid in the ensuing low temperature epoch where MSW conversions are able to enhance the asymmetry to values of order /0.2-0.37. DHPS do claim to find a significant final asymmetry for very large δm^2 values. However, for this regime the effective potential they employed is not valid.

Comments on the neutrino oscillation scene
Kayser, B.
Weak Interactions and Neutrinos, 2000/1/1, 339
Abstract - Not Available

Comments on the Physics Potential of Ultra-High Neutrino Interactions with Owl
Cline, D. B.
AIP Conf. Proc. 433: Workshop on Observing Giant Cosmic Ray Air Showers From >10(20) eV Particles From Space, 1998/1/1, 262
Abstract - Not Available

Comparative Analysis of GALLEX-GNO Neutrino Data and SXT X-Ray Data
Sturrock, Peter A.; Weber, M. A.
Multi-Wavelength Observations of Coronal Structure and Dynamics -- Yohkoh 10th Anniversary Meeting, 2001/1/1, 110
Abstract - There is some evidence that the low-energy component of the solar neutrino flux exhibits rotational modulation. The power spectrum of the GALLEX-GNO neutrino data has a peak at 13.59 +/- 0.06 y-1 (period = 26.88 +/- 0.12 days), that is within the band of synodic rotation rates of the solar convection zone. In order to relate the neutrino time series to the Sun's internal rotation, we have formed a "resonance statistic" that is a measure of the degree of "resonance" of the neutrino flux with the Sun's internal rotation, as determined by MDI. A map of this statistic indicates that the source of the modulation of the solar neutrino flux is located deep in the convection zone. We have now carried out a comparative analysis of GALLEX-GNO neutrino data and SXT X-ray data. We have formed the logarithm of the mean daily X-ray flux (to de-emphasize active regions) and then formed the mean power spectrum for five 15-degree latitude bands centered on 30S, 15S, the equator, 15N, and 30N. The mean spectrum peaks at 13.54 +/- 0.08 y-1 (period = 26.98 +/- 0.16 days). Taking account of the error bars, the neutrino peak frequency and SXT peak frequency are indistinguishable. We find that the neutrino and X-ray waveforms are approximately in anti-phase. We have also formed a map of the resonance statistic formed from the SXT power spectrum and the MDI rotation estimates. The neutrino-rotation and X-ray-rotation resonance maps are very similar, and both point to the same location deep in the convection zone. These results suggest that modulation of the low-energy solar neutrino flux and long-lived quasi-rigid rotation of the corona have a common cause. It seems possible that a magnetic structure deep in

the convection zone leads to (a) an enhancement of the X-ray flux, and (b) a diminution of the observed neutrino flux. This diminution could be caused by any one of several proposed mechanisms by which an electron neutrino, with non-zero magnetic moment and non-zero mass, may be converted into a neutrino of different flavor and/or opposite spin.

Comparative Analysis of GALLEX-GNO Neutrino Data, SOHO-MDI Helioseismology Data, and SXT X-ray Data
Sturrock, P. A.; Weber, M. A.
American Geophysical Union, Spring Meeting 2001,
Abstract - There have been many claims of correlation between solar neutrino measurements and various solar indices, including coronal brightness. These claims have been greeted with skepticism, largely on the grounds that there is no obvious mechanism to explain the correlation. We review our recent comparative analysis of GALLEX-GNO neutrino data and SOHO-MDI measurements of internal solar rotation, which yields evidence of rotational modulation of the solar neutrino flux. We attribute this modulation to the influence of internal magnetic structures on neutrino propogation. The coronal X-ray flux also exhibits rotational modulation. From this viewpoint, some kind of correlation between neutrino measurements and X-ray measurements is to be expected. We compare the spectra of the two time series with each other and with the distribuition of rotation frequencies in the solar interior. We also examine the correlation spectrum, which gives information about the frequencies responsible for correlation, and about the relative phases of the two oscillations. In order to relate these new studies with earlier studies (which did not involve spectrum analysis), we process the X-ray data in just the same way that the neutrino experiments process the solar neutrino flux. In this way we obtain a sequence of proxy X-ray measurements corresponding to the same start times, end times, and decay time, that apply to the GALLEX-

GNO runs. We can then carry out a simple correlation analysis of the neutrino and X-ray run sequences. We shall present the results of these studies and compare these results with earlier claims of correlation between neutrino measurements and solar indices. We wish to acknowledge support by NASA grants NAS 8-37334 and NAGW-2265 and NSF grant ATM-9910215.

Comparative Analysis of GALLEX-GNO Solar Neutrino Data and SOHO/MDI Helioseismology Data: Further Evidence for Rotational Modulation of the Solar Neutrino Flux
Sturrock, Peter A.; Weber, Mark A.
Astrophysical Journal, 565, 2002/2/1, 1366-1375
Abstract - Recent histogram analysis of GALLEX-GNO and SAGE data indicates that the solar neutrino flux, in the energy range of gallium experiments, varies on a timescale of weeks. Such variability could be caused by modulation of the neutrino flux by an inhomogeneous magnetic field in the solar interior if neutrinos have a nonzero magnetic moment. We may then expect the detected neutrino flux to oscillate with a frequency set by the synodic rotation frequency in the region of the solar interior that contains the magnetic structure. We investigate this possibility by carrying out a comparative analysis of the GALLEX-GNO solar neutrino data and estimates of the solar internal rotation rate derived from the MDI helioseismology experiment on the SOHO spacecraft. We find that while the Lomb-Scargle spectrum does not show a significant peak in the band appropriate to the radiative zone, it does show two closely spaced peaks in the band appropriate to the convection zone. In order to explore the relationship of these features to the Sun's internal rotation, we introduce a "resonance statistic" that is a measure of the degree of "resonance" of oscillations in the neutrino flux and the solar rotation as a function of radius and latitude. A two-dimensional map of the resonance statistic indicates that the modulation is occurring in the convection zone, near the equator. In order to derive a

significance estimate for this result, we next evaluate the integral of this statistic over selected equatorial sections corresponding to the convection zone and the radiative zone. This statistic yields strong evidence that modulation is occurring in the convection zone and no evidence that modulation is occurring in the radiative zone.

Comparison of Neutrino Transfer Methods for 1d Supernova Simulations
Suzuki, H.; Yamada, S.; Janka, H.-T.
Neutron Stars and Pulsars: Thirty Years after the Discovery, 1998/1/1, 75
Abstract - Not Available

Computing the collapse of iron-oxygen stellar cores with allowance for the absorption and emission of electron neutrinos and antineutrinos
Aksenov, A. G.
Astronomy Letters, 25, 1999/5/1, 307-317
Abstract - The collapse of a 1.4 M_solar iron stellar core and a 2 M_solar iron-oxygen core is computed by using 1D models. The P rho^{1+1/n} polytropes with n = 3 were chosen as the initial models. The equation of state takes into account photon equilibrium radiation, a mixture of Fermi gases of free nucleons and ideal gases of (Fe, He) nuclei in equilibrium relative to nuclear reactions, and an electron-positron gas. The problem includes the transfer equations for electron neutrinos and antineutrinos. We allow for the absorption and emission of neutrinos and antineutrinos which involve free nucleons and nuclei. The solution yielded neutrino light curves. The computed light curves exhibit narrow peaks with characteristic widths of approximately 10 ms. A constraint on the electron-neutrino mass (less than 4 eV) can thus be placed when short bursts of radiation are recorded during observations. Part of the energy of neutrino radiation is absorbed by the stellar core envelope: 3.6 x 10^{50} and 1.7 x 10^{50} erg for the 1.4 M_solar and 2 M_solar models, respectively. We also compute the collapse of a 2 M_solar stellar core with rapid initial rigid rotation with the averaging of the centrifugal force over the solid angle. The formation of a rapidly rotating neutron star in the final state points to the possibility of system fragmentation during the collapse.

Conditions for shock revival by neutrino heating in core-collapse supernovae
Janka, H.-Th.
Astronomy and Astrophysics, 368, 2001/3/1, 527-560
Abstract - Energy deposition by neutrinos can rejuvenate the stalled bounce shock and can provide the energy for the supernova explosion of a massive star. This neutrino-heating mechanism, though investigated by numerical simulations and analytic studies, is not finally accepted or proven as the trigger of the explosion. Part of the problem is that different groups have obtained seemingly discrepant results, and the complexity of the hydrodynamic models often hampers a clear and simple interpretation of the results. This demands a deeper theoretical understanding of the requirements of a successful shock revival. A toy model is developed here for discussing the neutrino heating phase analytically. The neutron star atmosphere between the neutrinosphere and the supernova shock can well be considered to be in hydrostatic equilibrium, with a layer of net neutrino cooling below the gain radius and a layer of net neutrino heating above. Since the mass infall rate to the shock is in general different from the rate at which gas is advected into the neutron star, the mass in the gain layer varies with time. Moreover, the gain layer receives additional energy input by neutrinos emitted from the neutrinosphere and the cooling layer. Therefore the determination of the shock evolution requires a time-dependent treatment. To this end the hydrodynamical equations of continuity and energy are integrated over the volume of the gain layer to obtain conservation laws for the total mass and energy in this layer. The radius and velocity of the supernova shock can then be calculated from global properties of the gain layer as solutions of an initial value problem, which expresses the fact that the behavior of the shock is controlled by the cumulative

effects of neutrino heating and mass accumulation in the gain layer. The described toy model produces steady-state accretion and mass outflow from the nascent neutron star as special cases. The approach is useful to illuminate the conditions that can lead to delayed explosions and in this sense supplements detailed numerical simulations. On grounds of the model developed here, a criterion is derived for the requirements of shock revival. It confirms the existence of a minimum neutrino luminosity that is needed for shock expansion, but also demonstrates the importance of a sufficiently large mass infall rate to the shock.

Confluence of cosmology, massive neutrinos, elementary particles, and gravitation
Kursunoglu, Behram N.; Mintz, Stephan L.; Perlmutter, Arnold
Confluence of Cosmology, Massive Neutrinos, Elementary Particles, and Gravitation, 1999/1/1
Abstract - Not Available

Connection Between Relic Neutrinos and Cosmic Rays at >= 10^{20} eV?
Weiler, T. J.
AIP Conf. Proc. 444: Particle Physics and Cosmology, First Tropical Workshop, 1998/1/1, 105
Abstract - Not Available

Constraining the window on sterile neutrinos as warm dark matter
Hansen, Steen H.; Lesgourgues, Julien; Pastor, Sergio; Silk, Joseph
Monthly Notices of the Royal Astronomical Society, 333, 2002/7/1, 544

Constraints on R-Parity Breaking in GUT Constrained Mssm from the Neutrinoless Double Beta Decay
Wodecki, A.; Kaminski, W. A.; Pagerka, S.
Particle and Nuclear Physics, 1998/1/1, 333
Abstract - Not Available

Contribution of the Weak Interaction to Neutrino Thermodynamics in

Astrophysical Plasmas
Tsintsadze, N. L. et al.
New Worlds in Astroparticle Physics II, 1999/1/1, 249
Abstract - Not Available

Correlative Aspects of the Solar Electron Neutrino Flux and Solar Activity
Wilson, Robert M.
Astrophysical Journal, 545, 2000/12/1, 532-546
Abstract - Between 1970 and 1994, the Homestake Solar Neutrino Detector obtained 108 observations of the solar electron neutrino flux (greater than 0.814 MeV). The "best fit" values derived from these observations suggest an average daily production rate of about 0.485 ^{37}Ar atom per day, a rate equivalent to about 2.6 SNU (solar neutrino units) or about a factor of 3 below the expected rate from the standard solar model. In order to explain, at least, a portion of this discrepancy, many researchers have speculated that the flux of solar neutrinos is variable, possibly being correlated with certain markers of the solar cycle (specifically, sunspot number and the Ap index). Indeed, previous studies, on the basis of shorter time intervals or data averaged in particular ways, often found evidence supportive for preferential behavior between the solar neutrino flux and solar activity. In this paper, using the larger "standard data set" and run-length-adjusted averages, the notion of preferential behavior between solar electron neutrino flux and solar activity is reexamined. The results clearly show that no statistically meaningful associations exist between the solar electron neutrino flux and any of the usual markers of solar activity, including sunspot number, the Ap index, the Deep River neutron monitor counts (cosmic rays), solar irradiance, and the number or size of solar energetic events (flares).

Cosmic microwave background anisotropy in the decaying neutrino cosmology
Adams, J. A.; Sarkar, S.; Sciama, D. W.
Monthly Notices of the Royal Astronomical Society, 301, 1998/11/1, 210-214

Abstract - It is attractive to suppose for several astrophysical reasons that the Universe has close to the critical density in light (~30 eV) neutrinos which decay radiatively with a lifetime of $\sim10^{23}$ s. In such a cosmology the Universe is re-ionized early and the last scattering surface of the cosmic microwave background significantly broadened. We calculate the resulting angular power spectrum of temperature fluctuations in the cosmic microwave background. As expected, the acoustic peaks are significantly damped relative to the standard case. This would allow a definitive test of the decaying neutrino cosmology with the forthcoming MAPPlanck surveyor missions.

Cosmic Neutrino Oscillations
Berezinsky, V.
Abstracts of the 19th Texas Symposium on Relativistic Astrophysics and Cosmology, held in Paris, France, Dec. 14-18, 1998. Eds.: J. Paul, T. Montmerle, and E. Aubourg (CEA Saclay)., 1998/12/1, 373
Abstract - The evidences for neutrino oscillations are found in atmospheric and solar neutrinos. The strongest evidence for atmospheric neutrino oscillations is given by zenith-angular dependence of muon neutrino flux. In case of solar neutrinos the evidence is weaker. All five solar-neutrino experiments show the deficit of the flux by a factor 2-3 as compared with the Standard Solar Model prediction. The latter is strongly confirmed by helioseismic measurements. Thus, solar neutrino measurements have a status of disappearance oscillation experiment. The atmospheric and solar neutrinos will be reviewed in the light of recent SuperKamiokande data. The consequences for other astrophysical phenomena will be discussed.

Cosmic Rays and Neutrinos from Gamma-Ray Bursts
Rachen, J. P.; Mészáros, P.
Gamma-Ray Bursts, 4th Hunstville Symposium, 1998/1/1, 776
Abstract - Not Available

Cosmic-ray neutrino annihilation on relic neutrinos revisited: a mechanism for generating air showers above the Greisen-Zatsepin-Kuzmin cutoff
Weiler, Thomas J.
Astroparticle Physics, 11, 1999/7/1, 303-316
Abstract - If neutrinos are a significant contributor to the matter density of the universe, then they should have ~ eV mass and cluster in galactic (super) cluster halos, and possibly in galactic halos as well. It was noted in the early 1980's that cosmic ray neutrinos with energy within $deltaE/E\_R=Gamma\_Z/M\_Z$ ~3% of the peak energy $E\_R=4(eV/m\_nu)x10^{21}$ eV will annihilate on the nonrelativistic relic antineutrinos (and vice versa) to produce the Z-boson with an enhanced, resonant cross section of $\ℴ(G\_F)\sim10^{-32}$ cm^2. The result of the resonant neutrino annihilation is a hadronic Z-burst 70% of the time, which contains, on average, thirty photons and 2.7 nucleons with energies near or above the GZK cutoff energy of $5x10^{19}$ eV. These photons and nucleons produced within our Supergalactic halo may easily propagate to earth and initiate super-GZK air showers. Here we show that the probability for each neutrino flavor at its resonant energy to annihilate within the halo of our Supergalactic cluster is likely within an order of magnitude of 1%, with the exact value depending on unknown aspects of neutrino mixing and relic neutrino clustering. The absolute lower bound in a hot Big Bang universe for the probability to annihilate within a 50 Mpc radius (roughly a nucleon propagation distance) of earth is 0.036%. From fragmentation data for Z-decay, we estimate that the nucleons are more energetic than the photons by a factor ~10. Several tests of the hypothesis are indicated.

Cosmological Limits on the Neutrino Mass from the LyAlpha Forest
Croft, R. A. C.; Hu, W.; Davé, R.
Physical Review Letters, 83, 1999/8/1, 1092
Abstract - The Lya forest in quasar spectra probes scales where massive neutrinos can

strongly suppress the growth of mass fluctuations. Using hydrodynamic simulations with massive neutrinos, we successfully test techniques developed to measure the mass power spectrum from the forest. A recent observational measurement in conjunction with a conservative implementation of other cosmological constraints places upper limits on the neutrino mass: m_{nu} < 5.5 eV for all values of $Omega_m$, and m_{nu} < 2.4 ($Omega_m$/0.17 -1) eV, if 0.2 < $Omega_m$ <0.5 as currently observationally favored (both 95 % C.L.).

Cosmological Neutrino Background Revisited
  Gnedin, Nickolay Y.; Gnedin, Oleg Y.
  Astrophysical Journal, 509, 1998/12/1, 11-15
  Abstract - We solve the Boltzmann equation for cosmological neutrinos around the epoch of the electron-positron annihilation in order to verify the freeze-out approximation and to compute accurately the cosmological neutrino distribution function. We find the radiation energy density to be about 0.3% higher than predicted by the freeze-out approximation. As a result, the spectrum of the cosmic microwave background anisotropies changes by ~0.3%-0.5%, depending on the angular scale, and the amplitude of the mass fluctuations on scales below about 100 h^-1 Mpc decreases by about 0.2%-0.3%.

Could intergalactic dust obscure a neutrino decay signature?
  Overduin, J. M.; Seahra, S. S.; Duley, W. W.; Wesson, P. S.
  Astronomy and Astrophysics, 349, 1999/9/1, 317-322
  Abstract - Following recent results from the Super-Kamiokande experiment which indicate that neutrinos may have finite masses and provide at least part of the dark matter in the Universe, we re-examine the decaying neutrino hypothesis of Sciama, including for the first time the effects of absorption by intergalactic dust. We consider several dust models, including one designed to optimize ultraviolet extinction for a given dust-to-gas ratio.

Dust absorption is insufficient to reconcile the theory with observational upper limits on the intensity of the diffuse ultraviolet background reported recently by Murthy et al. The theory remains marginally compatible with other diffuse background measurements.

CP, T violation in neutrino oscillations
  Bernabéu, J.
  Weak Interactions and Neutrinos, 2000/1/1, 227
  Abstract - Not Available

CP-Violation Effects in Long Baseline Neutrino Oscillation Experiments
  Koike, M.
  New Era in Neutrino Physics , 1998/1/1, 199
  Abstract - Not Available

CP-Violation in Neutrino Oscillations: Theory
  Dick, K.
  Proceedings of the Sixth SFB-375 Ringberg Workshop Astroteilchenphysik, 2000/2/1, 23
  Abstract - Not Available

Critical Solution in the Collapsing-Exploding Scenario of a Neutrino-Radiating Star
  González-Romero, L. M.
  Relativity and Gravitation in General, 1999/1/1, 263
  Abstract - Not Available

Cryogenic Detectors for Radiochemical Solar Neutrino Experiments
  Schnagl, J.
  Particle and Nuclear Physics, 1998/1/1, 147
  Abstract - Not Available

Current aspects of neutrino physics
  Caldwell, David O.
  Current aspects of neutrino physics, 2001/1/1
  Abstract - This book covers the currently most interesting aspects of neutrino physics written by leading experts of the field. The book starts with a history of neutrinos and then develops from the fundamentals to the direct determination of masses and lifetimes. The role of neutrinos

in fundamental astrophysical problems is discussed in detail. The book gives an up-to-date overview of modern neutrino physics and is useful for scientists and graduate students alike.

Current Status of Neutrino Oscillation Hunting
- CHORUS and NOMAD
Kodama, K.
Particles, Strings and Cosmology
(PASCOS 98), 1999/1/1, 272
Abstract - Not Available

Current status of the resonant spin-flavor precession solution to the solar neutrino problem
Guzzo, M. M.; Nunokawa, H.
Astroparticle Physics, 12, 1999/10/1, 87-95
Abstract - We discuss the current status of the resonant spin-flavor precession (RSFP) solution to the solar neutrino problem. We perform a fit to all the latest solar neutrino data for various assumed magnetic field profiles in the Sun. We show that the RSFP can account for all the solar neutrino experiments, giving as good a fit as other alternative solutions such as MSW or vacuum oscillation (Just So), and therefore can be a viable solution to the solar neutrino problem

Decaying Neutrinos and Large-Scale Structure Formation
Bharadwaj, Somnath; Sethi, Shiv K.
Astrophysical Journal Supplement Series, 114, 1998/1/1, 37
Abstract - We study the growth of density perturbations in a universe with unstable dark matter particles. The mass (m nu) range 30 eV <= m nu <= 10 keV with lifetimes (td) in the range 107 s <= td <= 1016 s are considered. We calculate the COBE normalized matter power spectrum for these models. We find that it is possible to construct models consistent with observations for masses m nu > 50 eV by adjusting td so as to keep the quantity $m^{2}_{nu}({keV})t_{d}({yr})$ constant at a value around 100. For m nu <= 1 keV, the power spectrum has extra power at small scales that could result in an early epoch of galaxy formation. We do not find any value of td that gives a viable

model in the mass range m nu <= 50 eV. We also consider the implications of radiatively decaying neutrinos--models in which a small fraction B << 1 of neutrinos decay into photons--that could possibly ionize the intergalactic medium (IGM) at high redshift. We show that the parameter space of decaying particles that satisfies the IGM observations does not give viable models of structure formation.

Decaying Neutrinos and the Flattering of the Galactic Halo
Sciama, Dennis W.
On Einstein's Path, essays in honor of Engelbert Schucking, 1999/1/1, 443
Abstract - Introduction The Neutrino Density Near the Sun τ^{23} and the Hα Data τ^{23} and the Extragalactic Background at 1500° A Conclusions

Degenerate and Other Neutrino Mass Scenarios and Dark Matter
Minakata, H.
Dark Matter in Astro- and Particle Physics, 2001/1/1, 404
Abstract - Not Available

Detecting Ultra High Energy Neutrinos by Upward Tau Airshowers and Gamma Flashes
Fargion, D.
Sources and Detection of Dark Matter and Dark Energy in the Universe, 2001/1/1, 516
Abstract - Not Available

Detection of Very Small Neutrino Masses in Double-Beta Decay Using Laser Tagging
Piepke, A.; Vogel, P.; Picchi, P.; Devoe, R.; Danilov, M.; Dolgolenko, A.; Zeldovich, O.; Vuilleumier, J.-L.; Pietropaolo, F.; Gratta, G.; Wang, Y.-F.; Giannini, G.
Identification of Dark Matter, 2001/1/1, 570
Abstract - Not Available

Determining the Supernova Direction by Its Neutrinos
Ando, S.; Sato, K.
Progress of Theoretical Physics, 107,

2002/5/1, 957-966
Abstract - Supernova neutrinos, which arrive at Earth earlier than light, allow for the earliest determination of the direction of the supernova. The topic of this paper is to study how accurately we can determine the supernova direction. We simulate supernova neutrino events at the SuperKamiokande detector, using a realistic supernova model and several realistic neutrino oscillation models. With the results of our simulation, we can restrict the supernova direction to be within a circle of radius 9 deg. In several neutrino oscillation models, this accuracy is increased to 8 deg. We also discuss the influence of an accident that occurred at the SuperKamiokande detector. After repair of the detector, using the remaining PMTs, the accuracy becomes about 12 deg for no oscillation.

Developing an Observatory for Multiflavor Neutrino Interactions from Supernovae (OMNIS)
Fenyves, E. J.
Abstracts of the 19th Texas Symposium on Relativistic Astrophysics and Cosmology, held in Paris, France, Dec. 14-18, 1998. Eds.: J. Paul, T. Montmerle, and E. Aubourg (CEA Saclay)., 1998/12/1, 278
Abstract - OMNIS is a proposed dedicated detector for mu and tau neutrinos from an SN in our Galaxy. The detector will be used to observe the time profile of about 1,000 events and would enable a non-zero mu or tau neutrino mass to be measured through the time-of-flight difference. This measurement, in conjunction with Super-Kamiokande and other detectors, will provide unique information on neutrino mixing effects over Galactic distances. Determination of the three neutrino masses has been of fundamental importance to particle physics and cosmology including the problem of Dark Matter.

Digital optical module and system design for km-scale neutrino detector in ice
Nygren, D.
New Astronomy Review, 42, 1998/9/1, 301-318
Abstract - Electronic Article Available from Elsevier Science.

Direct dark matter detection and neutrinoless double beta decay with an array of 40 kg of 'naked' natural Ge and 11 kg of enriched ^{76}Ge detectors in liquid nitrogen
Baudis, L.; Dietz, A.; Heusser, G.; Majorovits, B.; Strecker, H.; Klapdor-Kleingrothaus, H. V.
Astroparticle Physics, 17, 2002/6/1, 383-391
Abstract - Detection of the recoil energy deposited by a weakly interacting massive particle (WIMP) scattering off a nucleus or of the neutrinoless double beta decay signature in a 'naked' (natural or enriched) Ge crystal immersed in liquid nitrogen provides a new, yet simple implementation of a well know technology to the fields of direct dark matter and double beta decay searches. We show that an array with a total mass of 40 kg of natural Ge and 11 kg of enriched ^{76}Ge detectors operated in liquid nitrogen in a compact setup could yield important physics results by directly looking for a WIMP signature and testing the Majorana neutrino mass down to 0.1 eV. The method could be easily extended to much larger masses and, by increasing the amount of liquid nitrogen surrounding the detectors, to much lower backgrounds.

Direct measurements of neutrino mass
Wilkerson, J. F.; Robertson, R. G. H.
Current aspects of neutrino physics, 2001/1/1, 39
Abstract - Contents: 1. Introduction. 2. Overview of direct measurements. 3. Beta decay and electron capture measurements (ν_e). 4. Particle decay measurement - ν_μ and ν_τ. 5. Summary.

Direct search for mass of neutrino and anomaly in the tritium beta-spectrum
Lobashev, V. M. et al.
Weak Interactions and Neutrinos, 2000/1/1, 155
Abstract - Not Available

Discovering Ultra-High-Energy Neutrinos through Horizontal and Upward τ Air Showers: Evidence in Terrestrial Gamma Flashes?

Fargion, D.
Astrophysical Journal, 570, 2002/5/1, 909-925
Abstract - Ultra-high-energy (UHE) neutrinos ν_τ, ν_{τ}, and ν_e at PeV and higher energies may induce τ air showers whose detectability is amplified millions to billions of times by their secondaries. We considered UHE ν_τ -N and UHE ν_e-e interactions underneath mountains as a source of such horizontal amplified τ air showers. We also consider vertical upward UHE ν_τ - N interactions (UPTAUs) on Earth's crust, leading to UHE τ air showers or interactions at the horizon edges (HORTAUs), and their beaming toward high mountain gamma, X-ray, and Cerenkov detectors, and we show their detectability.

Do Statistically Significant Correlations Exist between the Homestake Solar Neutrino Data and Sunspots?
Boger, J.; Hahn, R. L.; Cumming, J. B.
Astrophysical Journal, 537, 2000/7/1, 1080-1085
Abstract - It has been suggested by various Authors that a significant anticorrelation exists between the Homestake solar neutrino data and the sunspot cycle. Some of these claims rest on smoothing the data by taking running averages, a method that has recently undergone criticism. We demonstrate that no significant anticorrelation can be found in the Homestake data, or in standard averages of that data. However, when running averages are taken, an anticorrelation seems to emerge whose significance grows as the number of points in the average increases. Our analysis indicates that the apparently high significance of these anticorrelations is an artifact of the failure to consider the loss of independence introduced in the running average process. When this is considered, the significance is reduced to that of the unaveraged data. Furthermore, when evaluated via parametric subsampling, no statistically significant anticorrelation is found. We conclude that the Homestake data cannot be used to substantiate any claim of an anticorrelation with the sunspot cycle.

Do We Understand the Origin and Impact of the Neutrino Flux in a Core Collapse Supernova?
Liebendoerfer, M.; Messer, O. E. B.; Mezzacappa, A.; Hix, W. R.; Bruenn, S. W.; Thielemann, F.-K.
American Astronomical Society Meeting, 200, 2002/5/1
Abstract - General relativistic multi group and multi flavor Boltzmann neutrino transport in spherical symmetry has become available - a new level of sophistication in the modeling of core collapse supernovae. We present postbounce simulations for five type II progenitors with 13-40 solar masses. Core collapse and bounce proceed in a quantitatively similar way, culminating in a rather uniform neutrino burst. At 100ms after bounce the shock stalls 150km radius in all our models. As the outer layers of the progenitors, where the density profiles are more different, fall into the protoneutron star, the neutrino luminosities reflect the individual mass accretion rates. Based on standard input physics, and without considering convection, high infall velocities prevent significant heating for a neutrino-driven supernova in general relativity. We acknowledge funding by the NSF under contract AST-9877130, the Oak Ridge National Laboratory, managed by UT-Batelle, LLC, for the U.S. Department of Energy under contract DE-AC05-00OR22725,the Joint Institute for Heavy Ion Research, the Swiss National Science Foundation under contract 20-61822.00, NASA under contract NAG5-8405, a DoE HENP PECASE Award, and the DoE HENP SciDAC Program. Our Simulations have been carried out on the NERSC Cray SV-1.

Dynamical QCD predictions for ultrahigh energy neutrino cross sections
Glück, M.; Kretzer, S.; Reya, E.
Astroparticle Physics, 11, 1999/7/1, 327-334
Abstract - Neutrino-nucleon total cross sections for neutrino energies up to

ultrahigh energies (UHE), $E\_nu = 10^{12}$ GeV, are evaluated within the framework of the dynamical (radiative) parton model. The expected uncertainties of these predictions do not exceed the level of about 20% at the highest energies where contributions of parton distributions in the yet unmeasured region around $x \sim = 10^{-8}$ to 10^{-9} are non-negligible. This is far more accurate than estimated uncertainties of about $2^{+/-1}$ due to fixed power law extrapolations of parton distributions to $x < 10^{-5}$ required for calculating UHE cosmic neutrino event rates.

EeV Neutrinos
Alvarez-Muñiz, J.; Zas, E.
New Worlds in Astroparticle Physics II,
1999/1/1, 159
Abstract - Not Available

Effect of Anisotropic Neutrino Radiation on
Supernova Explosion Energy
Shimizu, Tetsuya M.; Ebisuzaki,
Toshikazu; Sato, Katsuhiko; Yamada,
Shoichi
Astrophysical Journal, 552, 2001/5/1, 756-
781
Abstract - Since SN 1987A, many observations have indicated that supernova explosions are not spherical. The cause of the asymmetric explosion is still controversial (e.g., asymmetry in the envelope, the convective engine in the central core or in the proto-neutron star). In our previous study, anisotropic neutrino radiation has been proposed as an explanation for this asymmetry. In this paper we carried out a series of systematic multidimensional numerical simulations in order to investigate the effect of anisotropic neutrino radiation itself on the supernova explosion energy. The neutrino luminosity and the degree of anisotropy in neutrino radiation were assumed as input parameters, and the numerical results for various parameters were compared with each other. It was found that only a few percent of anisotropy in the neutrino emission distribution is sufficient to increase the explosion energy by a large factor. The explosion energy calculated so far in many supernova models has tended to be too short to explain the observation. Anisotropy of 10% in neutrino radiation roughly corresponds to an enhancement of 4% in total neutrino luminosity as far as the explosion energy is concerned. The increase in the explosion energy due to anisotropic neutrino radiation can be explained as follows. Anisotropically emitted neutrinos locally heat the supernova matter and revive a stalled shock wave in the direction of enhanced radiation. The expansion of the gas by the shock propagation results in a decrease in the neutrino cooling (emission) rate that rapidly decreases with the matter temperature. It is this suppression of energy loss that contributes largely to the increase in explosion energy. The efficiency of neutrino heating (absorption) itself is almost unchanged between anisotropic and spherical models with available energy fixed for neutrinos. In order for a stalled shock wave to revive, enhancement of the local intensity in the neutrino flux is of great importance, rather than that of the total neutrino luminosity over all the solid angle. It is first pointed out that such local neutrino heating is capable of triggering a supernova explosion. Anisotropic neutrino radiation is considered to be a plausible mechanism for a "successful" explosion other than the so far suggested "convective trigger."

Effect of Coulomb collisions on time variations
of the solar neutrino flux
Malyshkin, L.; Kulsrud, R.
Monthly Notices of the Royal
Astronomical Society, 316, 2000/8/1, 249-
266
Abstract - We consider the processes that might suppress the time variations in the solar neutrino flux produced by the radial motion of the Earth through the neutrino interference pattern. We calculate these time variations and the extent to which they are suppressed by Coulomb collisions of the neutrino-emitting nuclei. This is done for both the 0.862-MeV ^{7}Be neutrino line and the continuous neutrino spectrum, assuming a Gaussian energy response function of the neutrino detector. We find that the

collisional decoherence averages out the time variations for neutrino masses m22-m12>~10⁻⁷eV²ℰMeV3/2. A simple and clear physical picture of the time-dependent solar neutrino problem is presented and qualitative coherence criteria are discussed.

Effect of internal waves on the solar neutrino flux
Montalbán, J.; Schatzman, E.
Astronomy and Astrophysics, 351, 1999/11/1, 347-358
Abstract - The neutrino production rate strongly depends on the temperature, and a small variation of chemical composition in the central regions can imply a large variation of neutrino flux. For some values of their frequencies and wavenumbers, the internal waves e xcited at the bottom of the solar convective zone are able to propagate until the central region of the star, producing a macroscopic diffusive process (Press 1981, Schatzman & Montalbán 1995). In this paper we analyse the effect of this mixing mechanism on the neutrino flux production. For that purpose, we introduce some improvements in the treatment of non-adiabatic internal waves in the central region of the Sun. We use the amplitude of these oscillations to determine the macroscopic diffusion coefficient linked to non-adiabatic propagation of gravity waves. We analyse the chemical mixing in the solar center and its effect on the decrease of solar neutrino production. We also study how the diffusion coefficient depends on the source function of internal waves, and we show that the amplitude of the diffusion coefficient in the internal regions is strongly affected by the characteristics of the source function at the boundary convective/stratified regimes.

Effects of nuclear re-interactions in quasi-elastic neutrino-nucleus scattering
Bleve, C.; Co', G.; De Mitri, I.; Bernardini, P.; Mancarella, G.; Martello, D.; Surdo, A.
Astroparticle Physics, 16, 2001/11/1, 145-155
Abstract - The effects of nuclear re-interactions in the quasi-elastic neutrino-nucleus scattering are investigated with a phenomenological model. We found that the nuclear responses are lowered and that their maxima are shifted towards higher excitation energies. This is reflected on the total /ν-nucleus cross-section in a general reduction of about /15% for neutrino energies above 300 MeV.

Electromagnetic Neutrino Properties and Neutrino Oscillations in Electromagnetic Fields
Egorov, A. M.; Lobanov, A. E.; Studenikin, A. I.
New Worlds in Astroparticle Physics II, 1999/1/1, 153
Abstract - Not Available

Electron Neutrino Mass Measurement by Supernova Neutrino Bursts and Implications on Hot Dark Matter
Totani, T.
New Era in Neutrino Physics , 1998/1/1, 203
Abstract - Not Available

Electron neutrino mass measurements by supernova neutrino bursts and implications for hot dark matter.
Totani, T.
Physical Review Letters, 80, 1998/1/1, 2039-2042
Abstract - Not Available

Electroweak baryon number violation and constraints on left-handed Majorana neutrino mass
Sarkar, U.
Lepton and Baryon Number Violation in Particle Physics, neral Astrophysics and Cosmology, 1999/1/1, 361
Abstract - Not Available

Electroweak Measurements and Neutrino Oscillations: The NuTeV and BooNE Experiments
Shaevitz, M. H.
AIP Conf. Proc. 444: Particle Physics and Cosmology, First Tropical Workshop, 1998/1/1, 28
Abstract - Not Available

Element-Loaded Organic Scintillators for Neutron and Neutrino Physics
Brudanin, V. B.; Gundorin, N. A.; Filossofov, D. V.; Nemtchenok, I. B.; Smolnikov, A. A.; Vasiliev, S. I.; Bregadze, V. I.
Identification of Dark Matter, 2001/1/1, 626
Abstract - Not Available

Erratum to "Cosmic-ray neutrino annihilation on relic neutrinos revisitede: a mechanism for generating air showers above the Greisen-Zatsepin-Kuzmin cutoff
Weiler, T. J.
Astroparticle Physics, 12, 2000/1/1, 379-379
Abstract - Not Available

Evidence against the Sciama Model of Radiative Decay of Massive Neutrinos
Bowyer, Stuart; Korpela, Eric J.; Edelstein, Jerry; Lampton, Michael; Morales, Carmen; Pérez-Mercader, Juan; Gómez, José F.; Trapero, Joaquín
Astrophysical Journal, 526, 1999/11/1, 10-13
Abstract - We report on spectral observations of the night sky in the band around 900 Å where the emission line in the Sciama model of radiatively decaying massive neutrinos would be present. The data were obtained with a high-resolution, high-sensitivity spectrometer flown on the Spanish satellite MINISAT. The observed emission is far less intense than that expected in the Sciama model.

Evidence for electron neutrino flavor change through measurement of the (8)B solar neutrino flux at the Sudbury Neutrino Observatory
Neubauer, Mark Stephen Ph.D.
Thesis, 2001/11/1, 2
Abstract - The Sudbury Neutrino Observatory (SNO) is a water Cerenkov detector designed to study solar neutrinos. Using 1 kiloton of heavy water as the target and detection medium, SNO is able to separately determine the flux of electron neutrinos (ν$_e$) and the flux of all active neutrinos from the Sun by measuring the rate of charged current (CC) and neutral current (NC) interactions with deuterons. A comparison of these interaction rates allows for direct observation of solar neutrino oscillations. SNO can also search for oscillations by comparing the rate of CC and neutrino- electron elastic scattering (ES) events, since ES has both charged current and neutral current sensitivity.

Evidence for Neutrino Oscillation from Super-Kamiokande
Totsuka, Y.; The Super-KAMIOKANDE Collaboration
Abstracts of the 19th Texas Symposium on Relativistic Astrophysics and Cosmology, held in Paris, France, Dec. 14-18, 1998. Eds.: J. Paul, T. Montmerle, and E. Aubourg (CEA Saclay)., 1998/12/1, 616
Abstract - Super-Kamiokande is a 50,000 ton water Cherenkov detector for atmospheric, solar and supernova neutrinos as well as for a search of proton decay. Based on more than 700-day data of atmospheric neutrinos, we obtained the following results: (1)the observed ratio mu/e is significantly smaller than the expected one, (2)a strong deficit of upward-going mu compared to downward-going mu, (3)a clear east-west effect of e-like events consistent with expectation. These results are quantitatively reproduced with the assumption of nu_mu -----> nu_tau oscillation while the case for no oscillation is completely ruled out. We have analyzed more than 700-day solar-neutrino data with E_vis >= q 5.5 MeV. There is a strong indication that the electron E_vis distribution does not follow the one expected from the ^8B neutrinos. Its implication on the solar neutrino problem will be discussed.

Evidence for Neutrino Oscillation Observed in Super-Kamiokande
Totsuka, Y.
Frontiers Science Series 23: Black Holes and High Energy Astrophysics, 1998/1/1, 343
Abstract - Not Available

Evidence for Neutrino Oscillations from LSND: Past, Present and Future
Stancu, I.; The LSND Collaboration

The Identification of Dark Matter ,
1999/1/1, 584
Abstract - Not Available

Evidence for Neutrino Oscillations in LSND
Smith, D.
Dark Matter in Astro- and Particle Physics,
2001/1/1, 503
Abstract - Not Available

Exotic mechanisms for the neutrino masses
Berezhiani, Z.
Lepton and Baryon Number Violation in
Particle Physics, neral Astrophysics and
Cosmology, 1999/1/1, 147
Abstract - Not Available

Experimental Program on the Future CERN-
Gran Sasso Neutrino Beam-Line (NGS)
Migliozzi, P.
New Era in Neutrino Physics , 1998/1/1,
253
Abstract - Not Available

Experimental summary of the working group
on neutrino oscillation
Palladino, V.
Weak Interactions and Neutrinos,
2000/1/1, 336
Abstract - Not Available

Exploring the Neutrino-Driven Explosion
Mechanism of Massive Stars
Janka, H.-T.
American Astronomical Society Meeting,
200, 2002/5/1
Abstract - In this talk I attempt to present
an unprejudiced discussion of our present
understanding of neutrino-driven
explosions and the requirements for
achieving future progress in theoretical
work. Neutrinos undoubtedly play a crucial
role for the dynamics of stellar core
collapse and the evolution of the supernova
shock. They clearly dominate the
energetics of neutron star formation, as
splendidly confirmed by the neutrino
detection in connection with Supernova
1987A. Unfortunately, we do not have
observational support for the theoretical
conception that neutrino energy deposition
behind the shock is the ultimate cause for
the explosion of "typical" supernovae.
Numerical simulations and analytic work,
however, demonstrate that neutrino-driven
explosions can be obtained for suitable
conditions and reasonable combinations of
relevant parameters. But the outcome of
computer models depends sensitively on
changes of the input physics as well as
details of the numerical treatment.
Discussing the neutrino-driven mechanism
is therefore a quantitative problem, and
conclusive answers are jeopardized by
deficiencies in the numerical description
and our still incomplete knowledge of the
physical conditions and neutrino
interactions in dense matter. Refering to
results from a new generation of core-
collapse simulations which solve the
Boltzmann equation for the neutrino
transport, I will show that quantitatively
meaningful models require the inclusion of
general relativity and the accurate (but so
far grossly approximated) treatment of
neutrino reactions below the
neutrinosphere. Advancement towards a
standard supernova model requires multi-
dimensional calculations to account for the
effects of hydrodynamic instabilities. The
latter cannot only decide about the viability
of the neutrino-driven mechanism, but can
also account for the anisotropies observed
in the majority of ordinary massive star
explosions. Acknowledgments: This work
was supported by the
Sonderforschungsbereich 375 on
"Astroparticle Physics" of the Deutsche
Forschungsgemeinschaft and grants of
computer time on the CRAY T90 and
CRAY SV1ex of the John von Neumann
Institute for Computing in Jülich.

Extragalactic neutrino background from very
young pulsars surrounded by supernova
envelopes
Bednarek, W.
Astronomy and Astrophysics, 378,
2001/10/1, L49-L52
Abstract - We estimate the extragalactic
muon neutrino background which is
produced by hadrons injected by very
young pulsars at an early phase after
supernova explosion. It is assumed that
hadrons are accelerated in the pulsar wind
zone which is filled with thermal photons

captured below the expanding supernova envelope. In collisions with those thermal photons hadrons produce pions which decay into muon neutrinos. At a later time, muon neutrinos are also produced by the hadrons in collisions with matter of the expanding envelope. We show that extragalactic neutrino background predicted by such a model should be detectable by the planned 1 km^2> neutrino detector if a significant part of pulsars is born with periods shorter than ~ 10 ms. Since such population of pulsars is postulated by the recent models of production of extremely high energy cosmic rays, detection of neutrinos with predicted fluxes can be used as their observational test.

Extreme-Energy Cosmic Rays: Puzzles, Models, and Maybe Neutrinos
Weiler, T. J.
AIP Conf. Proc. 579: Radio Detection of High Energy Particles, 2001/1/1, 58
Abstract - Not Available

Extremely high energy neutrinos, neutrino hot dark matter, and the highest energy cosmic rays.
Yoshida, S.; Sigl, G.; Lee, S.
Physical Review Letters, 81, 1998/1/1, 5505-5508
Abstract - Not Available

Field-theoretical treatment of neutrino oscillations
Grimus, W. et al.
Weak Interactions and Neutrinos, 2000/1/1, 355
Abstract - Not Available

Final Results of the Gallex Solar Neutrino Experiment
Vignaud, D.
New Worlds in Astroparticle Physics II, 1999/1/1, 117
Abstract - Not Available

First measurement of the flux of solar neutrinos from the sun at the Sudbury Neutrino Observatory
Wittich, Peter Ph.D.
Thesis, 2000/12/1, 8
Abstract - The Sudbury Neutrino Observatory (SNO) is a second generation solar neutrino detector. SNO is the first experiment that is able to measure both the electron neutrino flux and a flavor-blind flux of all active neutrino types, allowing a model-independent determination if the deficit of solar neutrinos known as the solar neutrino problem is due to neutrino oscillation. The Sudbury Neutrino Observatory started taking production data in November, 1999. A measurement of the charged current rate will be the first indication if SNO too sees a suppression of the solar neutrino signal relative to the theoretical predictions. Such a confirmation is the first step in SNO's ambitious science program. In this thesis, we present evidence that SNO is seeing solar neutrinos and a preliminary ratio of the measured vs predicted rate of electrons as induced by ^{8}B neutrinos in the ν$_e$, + d --> p + p + e charged-current (CC) reaction.

First Neutrino Observations from the Sudbury Neutrino Observatory
Sinclair, D.
Dark Matter in Astro- and Particle Physics, 2001/1/1, 493
Abstract - Not Available

First results from the CHOOZ experiment on neutrino oscillations
Giannini, G.; Apollonia, M.
Dark matter : proceedings of DM97, 1st Italian conference on dark matter, Trieste, December 9-11, 1997 / editor, Paolo Salucci. Firenze, Italy : Studio Editoriale Fiorentino, c1998, p. 79., 1998/1/1, 79
Abstract - Not Available

Flavor changing neutrino oscillations in LSND
White, D. H.
Weak Interactions and Neutrinos, 2000/1/1, 391
Abstract - Not Available

Flavor Physics: Kaons, Charm, Beauty, Taus, and Neutrinos
McBride, P. L.
AIP Conf. Proc. 423: Fundamental Particles and Interactions, Frontiers in

Contemporary Physics, 1998/1/1, 226
Abstract - Not Available

Flavour models for neutrino physics
Romanino, A.
Proceedings of the Sixth SFB-375
Ringberg Workshop Astroteilchenphysik,
2000/2/1, 33
Abstract - Not Available

Four-neutrino mixing and Big-Bang
Nucleosynthesis
Bilenky, S. M.; Giunti, C.; Grimus, W.;
Schwetz, T.
Astroparticle Physics, 11, 1999/9/1, 413-
428
Abstract - We investigate the Big-Bang
Nucleosynthesis constraints on the
elements of the neutrino mixing matrix
which connect sterile with massive
neutrino fields, in the framework of the
two four-neutrino schemes that are favored
by the results of neutrino oscillation
experiments. We discuss the implications
of these constraints for terrestrial short and
long-baseline neutrino oscillation
experiments and we present several
possibilities of testing them in these
experiments. In particular, we show that
from the Big-Bang Nucleosynthesis
constraints it follows that the nu(-)_-3pt
mu -->nu(-)_-3pt tau transition is
severely suppressed in short-baseline
experiments, whereas its oscillation
amplitude in long-baseline experiments is
of order 1. We also propose a new
parameterization of the four-neutrino
mixing matrix U which is appropriate for
the schemes under consideration.

Four-Neutrino Mixing, Oscillations and Big-
Bang Nucleosynthesis
Bilenky, S. M.; Giunti, C.; Grimus, W.;
Schwetz, T.
New Era in Neutrino Physics , 1998/1/1,
179
Abstract - Not Available

Four-Neutrino Oscillations
Barger, V.
Particles, Strings and Cosmology
(PASCOS 98), 1999/1/1, 288

Four-neutrino spectrum from oscillation data
Bilenky, S. M.; Giunti, C.; Grimus, W.;
Schwetz, T.
Weak Interactions and Neutrinos,
2000/1/1, 195
Abstract - Not Available

Freezeout and Neutrinos in R-Process
Nucleosynthesis
Surman, Rebecca Ann Ph.D.
Thesis, 1998/1/1, 25
Abstract - The r process is the rapid
neutron capture nucleosynthesis process
responsible for the formation of
approximately half of the elements with A
> 70. While the main characteristics of
the r-process are understood, the
astrophysical site where it occurs has yet to
be pinned down. Here several aspects of
the r-process are investigated with the aim
of restricting further the conditions
necessary for a successful r process. The
rare earth peak, a previously poorly
understood feature in the r-process
abundance distribution, is studied and is
found to form as the very neutron-rich r-
process products decay to stability. This
understanding helps to constrain the
conditions under which the r process
freezes out. Further constraints are found
by investigating the role of neutrinos in the
r process. Neutrino-nucleus interaction
rates are calculated for select unstable
neutron-rich nuclei and are found to be
significantly larger than previous
estimates. The influence of these rates on
the r process strictly limits the neutrino
flux that can be present during a successful
r process.

Further Evidence for Neutrino Oscillations
from Lsnd: the ν_{μ} -->
ν_{e} Decay-In Channel
Stancu, I.; LSND Collaboration
Particle and Nuclear Physics, 1998/1/1,
167
Abstract - Not Available

Further studies of the OMNIS supernova
neutrino observatory: optimisation of
detector configuration and possible
extension to solar neutrinos
Smith, P. F.

Astroparticle Physics, 16, 2001/10/1, 75-96
Abstract - The basic design principles and earlier origins of OMNIS (Observatory for Multiflavour Neutrino Interactions from Supernovae) have been described in a previous paper [Astropart. Phys. 8 (1997) 27]. Its purpose would be to record a large number of mu and tau neutrinos from a supernova burst, complementing other world detectors which observe mainly electron antineutrinos. This would enable a cosmologically significant neutrino mass to be measured or definitely excluded by the arrival time profile. The detector is based on neutral current excitation of lead and iron target nuclei followed by release of neutrons. In this paper we summarise the published neutron production estimates for different targets, with and without mixing, and discuss the results of simulations of a range of target/detector configurations, with the objective of optimising the single and double neutron signals from a given target mass. Discussions are included of the choice of neutron detection method, and the effect of neutron and gamma backgrounds. It is further proposed that OMNIS detector modules might be designed to include real-time solar neutrino spectroscopy using the nuclear excitation principle devised by Raghavan [Phys. Rev. Lett. 78 (1997) 3618].

Future Dark Matter and Neutrino Physics Experiments at the UK Boulby Mine
Smith, P. F.
The Identification of Dark Matter ,
1999/1/1, 613
Abstract - Not Available

Future Detection of Supernova Neutrino Burst and Explosion Mechanism
Totani, T.; Sato, K.; Dalhed, H. E.; Wilson, J. R.
Astrophysical Journal, 496, 1998/3/1, 216
Abstract - Future detection of a supernova neutrino burst by large underground detectors would give important information for the explosion mechanism of collapse-driven supernovae. We studied the statistical analysis for the future detection of a nearby supernova by using a numerical supernova model and realistic Monte Carlo simulations of detection by the Super-Kamiokande detector. We mainly discuss the detectability of the signatures of the delayed explosion mechanism in the time evolution of the $\overline{\nu}_{e}$ luminosity and spectrum.

Future High Energy Neutrino Program at CERN
Palladino, V.
Particle and Nuclear Physics, 1998/1/1, 443
Abstract - Not Available

Future Neutrino Astrophysics Projects at the UK Boulby Mine
Smith, P. F. et al.
Sources and Detection of Dark Matter and Dark Energy in the Universe, 2001/1/1, 532
Abstract - Not Available

Future Prospects for Accelerator-Based Neutrino Mass Searches
Santacesaria, R.
The Identification of Dark Matter ,
1999/1/1, 612
Abstract - Not Available

Future solar neutrino projects: Borexino
Caccianiga, B.
Weak Interactions and Neutrinos, 2000/1/1, 365
Abstract - Not Available

Galactic gamma halo by heavy neutrino annihilations ?
Fargion, D.
Astroparticle Physics, 12, 2000/1/1, 307-314
Abstract - Not Available

Galactic Gamma Halo by Heavy Neutrino Annihilations?
Fargion, D.; Konoplich, R.; Grossi, M.; Khlopov, M.
The Identification of Dark Matter ,
1999/1/1, 500
Abstract - Not Available

Galactic rotational velocities and dark neutrino matter

Marx, G.; Kupi, G.
Weak Interactions and Neutrinos,
2000/1/1, 269
Abstract - Not Available

GALLEX Solar Neutrino Results
Kirsten, T. A.
Particle and Nuclear Physics, 1998/1/1, 85
Abstract - Not Available

Gamma-ray and neutrino flares produced by
protons accelerated on an accretion disc
surface in active galactic nuclei
Bednarek, W.; Protheroe, R. J.
Monthly Notices of the Royal
Astronomical Society, 302, 1999/1/1, 373-
380
Abstract - We discuss the consequences of
almost rectilinear acceleration of protons to
extremely high energies in a reconnection
region on the surface of an accretion disc
that surrounds a central black hole in an
active galaxy. The protons produce
gamma-rays and neutrinos in interactions
with the disc radiation as considered in
several previous papers. However, in this
model the secondary gamma-rays can
initiate cascades in the magnetic and
radiation fields above the disc. We
compute the spectra of gamma-rays and
neutrinos emerging from regions close to
the disc surface. Depending on the
parameters of the reconnection regions,
this model predicts the appearance of
gamma-ray and neutrino flares if protons
take most of the energy from the
reconnection region. In contrast, if leptons
take most of the energy, they produce pure
gamma-ray flares. The gamma-ray
spectrum expected in the case of hadronic
cascading is compared with the spectrum
observed during the flare in 1991 June
from 3C 279. The neutrino flares that
should accompany these gamma-ray flares
may be detected by future large-scale
neutrino telescopes sensitive at ~ 10^5
TeV.

Gamma-Ray Bursts via the Neutrino Emission
from Heated Neutron Stars
Salmonson, Jay D.; Wilson, James R.;
Mathews, Grant J.
Astrophysical Journal, 553, 2001/6/1, 471-
487
Abstract - A model is proposed for
gamma-ray bursts based upon a neutrino
burst of $\sim 10^{52}$ ergs lasting a few seconds
above a heated collapsing neutron star.
This type of thermal neutrino burst is
suggested by relativistic hydrodynamic
studies of the compression, heating, and
collapse of close binary neutron stars as
they approach their last stable orbit but
may arise from other sources as well. We
present a spherically symmetric
hydrodynamic simulation of the formation
and evolution of the pair plasma associated
with such a neutrino burst. This pair
plasma leads to the production of $\sim 10^{51}$-
10^{52} ergs in gamma rays with spectral and
temporal properties consistent with many
observed gamma-ray bursts.

Gamma-Ray Bursts: Afterglow, High-Energy
Cosmic Rays, and Neutrinos
Waxman, Eli
Astrophysical Journal Supplement Series,
127, 2000/4/1, 519-526
Abstract - The recent discovery of delayed,
low-energy emission, "afterglow," from
gamma-ray burst (GRB) sources confirmed
the cosmological origin of the bursts and
provided support for models in which
GRBs are produced by the dissipation of
the kinetic energy of relativistic fireballs.
In this talk, I review the fireball model,
outline the major open questions associated
with GRB fireball physics, and discuss
several phenomena that may be associated
with cosmological fireballs: GRB
afterglows, ultra-high-energy (>10^{19}
eV) cosmic rays, and high-energy (~10^{14})
neutrinos. Invited talk, Second
International Workshop on Laboratory
Astrophysics with Intense Lasers
(University of Arizona, Tucson, 1998
March).

Gamma-rays and neutrinos from very young
supernova remnants
Protheroe, R. J.; Bednarek, W.; Luo, Q.
Astroparticle Physics, 9, 1998/6/1, 1-14
Abstract - We consider the result of
acceleration of heavy ions in the slot gap
potential of a very young pulsar with a hot
polar cap. Photodisintegration of the heavy

ions in the radiation field of the polar cap and pulsar surface gives rise to a flux of energetic neutrons. Some fraction of these neutrons interact with target nuclei in the supernova shell to produce neutrino and gamma-ray signals which should be observable from very young supernova remnants in our galaxy for a range of pulsar parameters.

General Relativistic Augmentation of Neutrino Pair Annihilation Energy Deposition near Neutron Stars
Salmonson, Jay D.; Wilson, James R.
Astrophysical Journal, 517, 1999/6/1, 859-865
Abstract - General relativistic calculations are made of neutrino-antineutrino annihilation into electron-positron pairs near the surface of a neutron star. It is found that the efficiency of this process is enhanced over the Newtonian values up to a factor of more than 4 in the regime applicable to Type II supernovae, and by up to a factor of 30 for collapsing neutron stars.

General Relativistic Effects on Neutrino-driven Winds from Young, Hot Neutron Stars and r-Process Nucleosynthesis
Otsuki, Kaori; Tagoshi, Hideyuki; Kajino, Toshitaka; Wanajo, Shinya
Astrophysical Journal, 533, 2000/4/1, 424-439
Abstract - Neutrino-driven winds from young hot neutron stars, which are formed by supernova explosions, are the most promising candidate site for r-process nucleosynthesis. We study general relativistic effects on this wind in Schwarzschild geometry in order to look for suitable conditions for successful r-process nucleosynthesis. It is quantitatively demonstrated that general relativistic effects play a significant role in increasing the entropy and decreasing the dynamic timescale of the neutrino-driven wind. Exploring the wide parameter region that determines the expansion dynamics of the wind, we find interesting physical conditions that lead to successful r-process nucleosynthesis. The conditions that we found are realized in a neutrino-driven wind with a very short dynamic timescale, $\tau_{dyn} \sim 6$ ms, and a relatively low entropy, S~140. We carry out α-process and r-process nucleosynthesis calculations on these conditions with our single network code, which includes over 3000 isotopes, and confirm quantitatively that the second and third r-process abundance peaks are produced in neutrino-driven winds.

General Relativistic Simulations of Stellar Core Collapse and Postbounce Evolution with Boltzmann Neutrino Transport
Liebendörfer, M.; Messer, O. E. B.; Mezzacappa, A.; Hix, W. R.
20th Texas Symposium on relativistic astrophysics, 2001/1/1, 472
Abstract - Not Available

Generation of neutrino masses and mixings in gauge theories
Tanimoto, M.
Weak Interactions and Neutrinos, 2000/1/1, 345
Abstract - Not Available

GENIUS and the Genius TF: A New Observatory for WIMP Dark Matter and Neutrinoless Double Beta Decay
Klapdor-Kleingrothaus, H. V.; Majorovits, B.
Identification of Dark Matter, 2001/1/1, 593
Abstract - Not Available

Going Beyond the Salam-Weinberg Standard Model with Decaying Neutrinos
Sciama, Dennis W.
The Abdus Salam Memorial Meeting, 1999/1/1, 30
Abstract - Not Available

Grand Observatories and Multiple-OWL for High-Energy Neutrino Astrophysics
Takahashi, Y.; Dimmock, J. O.; Hillman, L. W.; Hadaway, J. B.; Lamb, D. J.; Mohri, M.; Ebisuzaki, T.
AIP Conf. Proc: 458: Space Technology and Applications, International Forum -- 1999, 1999/1/1, 315
Abstract - Not Available

Gravitational Phase Transition of Heavy
    Neutrino Matter
    Bilic, N.; Tsiklauri, D.; Viollier, R. D.
    Particle and Nuclear Physics, 1998/1/1, 17
    Abstract - Not Available

GRB990705: no evidence for neutrino signal.
    Fulgione, W.
    GRB Circular Network, 390, 1999/1/1, 1
    Abstract - Not Available

Green's Functions for Neutrino Mixing
    Blasone, M.; Vitiello, G.; Henning, P.
    Particles, Strings and Cosmology
    (PASCOS 98), 1999/1/1, 306
    Abstract - Not Available

Heavy neutrino ball as a possible solution to the
    "blackness problem" of the Galactic center
    Tsiklauri, David; Viollier, Raoul D.
    Astroparticle Physics, 12, 1999/11/1, 199-
    205
    Abstract - It has been recently shown
    (Tsiklauri and Viollier, 1998, ApJ 500,
    591) that the matter concentration inferred
    from observed stellar motion at the galactic
    center (Eckart and Genzel 1997, MNRAS
    284, 576 and Genzel et al. 1996, ApJ 472,
    153) is consistent with a supermassive
    object of 2.5 x10^6 solar masses,
    composed of self-gravitating, degenerate
    heavy neutrinos. It has been furthermore
    suggested (Tsiklauri and Viollier, 1998,
    ApJ 500, 591) that the neutrino ball
    scenario may have an advantage that it
    could possibly explain the so-called
    "blackness problem" of the galactic center.
    We present a quantitative investigation of
    this statement, by calculating the emitted
    spectrum of Sgr A^* in the framework of
    standard accretion disk theory.

Helioseismic Constraints on Solar Structure and
    the Solar Neutrino Problem
    Roxburgh, Ian W.
    Astrophysics and Space Science, 261,
    1998/1/1, 57-58
    Abstract - Not Available

Helioseismic constraints on the solar neutrino
    fluxes
    Vauclair, Sylvie
    Proceedings of the SOHO 10/GONG 2000

Workshop: Helio- and asteroseismology at
    the dawn of the millennium, 2-6 October
    2000, Santa Cruz de Tenerife, Tenerife,
    Spain. Edited by A. Wilson, Scientific
    coordination by P. L. Pallé. ESA SP-464,
    Noordwijk: ESA Publications Division,
    ISBN 92-9092-697-X, 2001, p. 491 - 496,
    464, 2001/1/1, 491
    Abstract - Solar models are more and more
    constrained by helioseismology.
    Meanwhile, solar neutrino measurements
    at the earth are more and more precise
    owing to the various detectors working
    simultaneously. The astrophysical
    solutions to the solar neutrino problem
    which have been proposed in the past are
    ruled out and neutrino oscillations seem the
    most promising explanation for the
    observed spectrum. However the situation
    is not yet completely settled, either in
    astrophysics or in particle physics.

Helioseismology and solar neutrinos
    Chitre, S. M.
    Bulletin of the Astronomical Society of
    India, 28, 2000/3/1, 71
    Abstract - Not Available

Helioseismology and Solar Neutrinos
    Christensen-Dalsgaard, J.
    Eighteenth Texas Symposium on
    Relativistic Astrophysics, 1998/1/1, 694
    Abstract - Not Available

Helioseismology and the solar neutrino
    problem
    Antia, H.; Chitre, S.
    IAU Symp. 185: New Eyes to See Inside
    the Sun and Stars, 185, 1998/1/1, 41
    Abstract - The accurately measured
    frequencies of solar oscillations can be
    inverted to determine the profiles of sound
    speed and density through a large part of
    the Sun's interior. This acoustic structure
    can be used to obtain the temperature and
    chemical composition profiles inside the
    Sun and also to calculate the expected
    neutrino fluxes. In the framework of
    standard neutrino physics, but with the
    allowance of arbitrary variations in the
    input opacities and even relaxation of the
    thermal equilibrium condition, it turns out
    to be difficult to produce a seismic model

that is simultaneously consistent with any two of the existing solar neutrino experiments. It is therefore tempting to suggest that the low observed fluxes of solar neutrinos should be attributed to nonstandard neutrino physics.

Hidden source of high-energy neutrinos in
    collapsing galactic nucleus
    Berezinsky, V. S.; Dokuchaev, V. I.
    Astroparticle Physics, 15, 2001/3/1, 87-96
    Abstract - We propose a model of a short-lived very powerful source of high energy (HE) neutrinos. It is formed as a result of the dynamical evolution of a galactic nucleus prior to its collapse into a massive black hole and formation of high-luminosity AGN. This stage can be referred to as "pre-AGN". A dense central stellar cluster in the galactic nucleus at the late stage of evolution consists of compact stars (neutron stars and stellar mass black holes). This cluster is sunk deep into the massive gas envelope produced by destructive collisions of a primary stellar population. Frequent collisions of neutron stars in a central stellar cluster are accompanied by the generation of ultrarelativistic fireballs and shock waves. These repeating fireballs result in a formation of the expanding rarefied cavity inside the envelope. The charged particles are effectively accelerated in the cavity and, due to pp collisions in the gas envelope, they produce HE neutrinos. All HE particles, except neutrinos, are absorbed in the thick envelope.

High Energy Cosmic Neutrino Astronomy: The
    ANTARES Project
    Basa, S.; The ANTARES Collaboration
    Abstracts of the 19th Texas Symposium on Relativistic Astrophysics and Cosmology, held in Paris, France, Dec. 14-18, 1998. Eds.: J. Paul, T. Montmerle, and E. Aubourg (CEA Saclay)., 1998/12/1, 619
    Abstract - Neutrinos should offer the unique opportunity to explore the Universe at ultra high energy and at long distance. The ANTARES Collaboration aims at demonstrating with a scalable prototype the feasability of an undersea neutrinos telescope and at developping and operating

mooring lines to select the optimal site to install a kilometer-scale detector. After a description of the expected sensitivity of such a detector to astrophysical sources (AGNs, X ray binary systems, young supernova remnants,...) and also to particle physics sources (supersymmetric particle decays, neutrinos oscillation, ...), the status of the achievements and of the future developments will be reported.

High energy cosmic-rays and neutrinos from
    cosmological gamma-ray burst fireballs.
    Waxman, E.
    Physica Scripta Volume T, 85, 2000/1/1, 117-126
    Abstract - Not Available

High energy neutrino astronomy
    Halzen, F.
    Weak Interactions and Neutrinos, 2000/1/1, 123
    Abstract - Not Available

High Energy Neutrino Astronomy and the
    ANTARES Project
    Montanet, F.
    25th International Winter Meeting on Fundamental Physics. Selected Topics on High Energy and Astropartical Physics., 1998/1/1, 338
    Abstract - Not Available

High Energy Neutrino Astronomy with Antares
    Moorhead, M. E.
    COSMO-97, First International Workshop on Particle Physics and the Early Universe, 1998/1/1, 115
    Abstract - Not Available

High Energy Neutrino Astronomy: PASCOS 99
    Halzen, Francis
    Particles, Strings and Cosmology, 2000/1/1, 349
    Abstract - Not Available

High Energy Neutrino Astronomy: Towards
    Kilometer-Scale Detectors
    Halzen, F.
    Current Topics in Astrofundamental Physics: the Cosmic Microwave Background, 2001/1/1, 585

High energy neutrino astronomy: towards
kilometer-scale detectors
Halzen, Francis
Current aspects of neutrino physics,
2001/1/1, 309
Abstract - Contents: 1. Introduction. 2.
Scientific goals - galactic sources,
extragalactic sources, particle physics,
other science. 3. Large natural Cerenkov
detectors.

High-energy astrophysics neutrinos at MACRO
di Credico, A.; The MACRO Collaboration
Weak Interactions and Neutrinos,
2000/1/1, 248
Abstract - Not Available

High-Energy Gamma and Neutrino Astronomy
Bergström, L.
Eighteenth Texas Symposium on
Relativistic Astrophysics, 1998/1/1, 58
Abstract - Not Available

High-Energy Neutrino Astronomy: Towards
Kilometer-Scale Detectors
Halzen, F.
High Energy Gamma-Ray Astronomy,
2001/1/1, 43
Abstract - Not Available

High-Energy Neutrinos from Photomeson
Processes in Blazars
Atoyan, Armen; Dermer, Charles D.
Physical Review Letters, 87, 2001/11/1,
1102
Abstract - Not Available

High-energy physics: Neutrinos reveal split
personalities
Bahcall, John N.
Nature, 412, 2001/7/1, 29-31
Abstract - Not Available

Histogram Analysis of GALLEX, GNO and
SAGE Neutrino Data: Further Evidence for
Variability of the Solar Neutrino Flux
Sturrock, P. A.; Scargle, J. D.
American Astronomical Society Meeting,
197, 2000/12/1
Abstract - If the solar neutrino flux were
constant, as is widely assumed, the
histogram of flux measurements would be
unimodal. On the other hand, sinusoidal or
square-wave modulation may lead to a
bimodal histogram. We here present
evidence that the neutrino flux histogram is
in fact bimodal. We analyze all available
data from gallium experiments,
coordinating results from the GALLEX
and GNO experiments into one data set,
and adopting results from the SAGE
experiment as another data set. The two
histograms, from the two data sets, are
consistent in showing peaks in the range 45
- 75 SNU and 90 - 120 SNU, and a valley
in between. By combining the data into one
data set, we may form more detailed
histograms; these strengthen the case that
the flux is bimodal. We wish to
acknowledge support (for PAS) by NASA
grants NAS 8-37334 and NAGW-2265 and
NSF grant ATM-9910215 and (for JDS) by
the NASA Applied Information Systems
Research Program.

Histogram Analysis of GALLEX, GNO, and
SAGE Neutrino Data: Further Evidence for
Variability of the Solar Neutrino Flux
Sturrock, Peter A.; Scargle, Jeffrey D.
Astrophysical Journal, 550, 2001/3/1,
L101-L104
Abstract - If the solar neutrino flux were
constant, as is widely assumed, the
histogram of flux measurements would be
unimodal. On the other hand, sinusoidal or
square-wave modulation (either periodic or
stochastic) may lead to a bimodal
histogram. We here present evidence that
the neutrino flux histogram is in fact
bimodal. We analyze all available data
from gallium experiments, coordinating
results from the GALLium EXperiment
and the Gallium Neutrino Observatory
experiment into one data set and adopting
results from the Soviet-American Gallium
Experiment as another data set. The two
histograms, from the two data sets, are
consistent in showing peaks in the ranges
45-75 and 90-120 SNU, with a valley in
between. By combining the data into one
data set, we may form more detailed
histograms; these strengthen the case that
the flux is bimodal. A preliminary
statistical analysis indicates that the
bimodal character of the solar neutrino flux
is highly significant. Since the upper peak

is close to the expected flux (120-140 SNU), we may infer that the neutrino deficit is due to time-varying attenuation of the flux produced in the core. We estimate the timescale of this variation to be in the range 10-60 days. Attenuation that varies on such a timescale is suggestive of the influence of solar rotation and points toward a process involving the solar magnetic field in conjunction with a nonzero neutrino magnetic moment.

How Uncertain are Solar Neutrino Predictions?
Bahcall, J. N.; Basu, S.; Pinsonneault, M. H.
Physics Letters B, 433, 1998/1/1, 128
Abstract - Not Available

Hydrodynamical Study of Neutrino-Driven Wind as an r-Process Site
Sumiyoshi, Kohsuke; Suzuki, Hideyuki; Otsuki, Kaori; Terasawa, Mariko; Yamada, Shoichi
Publications of the Astronomical Society of Japan, 52, 2000/8/1, 601-611
Abstract - We discuss the neutrino-driven wind from a proto-neutron star based on general-relativistic hydro-dynamical simulations. We examine the properties of the neutrino-driven wind to explore the possibility of r-process nucleosynthesis. Numerical simulations involving neutrino heating and cooling processes were performed with the assumption of a constant neutrino luminosity by using realistic profiles of the proto-neutron star (PNS) as well as simplified models. The dependence on the mass of PNS and the neutrino luminosity is presented systematically. Comparisons with analytic treatments in the previous studies are also given. In the cases with the realistic PNS, we have found that the entropy per baryon and the expansion time scale are neither high nor short enough for the r-process within the current assumptions. On the other hand, we have also found that the expansion time scale obtained by hydrodynamical simulations is systematically shorter than that in the analytic solutions due to our proper treatment of the equation of state. This fact might lead to an increase in the neutron-to-seed ratio, which is suitable for the r-process in a neutrino-driven wind.

ICECUBE: The Future of Neutrino Astronomy
Halzen, F.
American Astronomical Society Meeting, 192, 1998/5/1
Abstract - With an effective telescope area of order 10,000 meter squared, the AMANDA Antarctic Muon and Neutrino Detector Array represents the first of a new generation of high energy neutrino telescopes, reaching a scale envisaged over 25 years ago. We will argue however that a high energy neutrino telescope should have an effective area of order 1 square kilometer in order to detect neutrino emission from the most energetic cosmic processes involving pulsars, black holes, active galactic nuclei and the like. Such an instrument also has unique capabilities in searching for neutrino mass and dark matter. An international collaboration has recently taken the initiative to construct such a telescope by instrumenting one kilometer cubed of Antarctic ice as a neutrino detector.

II. Neutrino Masses and Neutrino Oscillations
di Lella, L.
Electroweak Physics, 2000/1/1, 84
Abstract - Not Available

Implications of Neutrino Flux Bounds for Future Neutrino Observations
Hettlage, C.; Mannheim, K.
Astronomische Gesellschaft Meeting Abstracts, 18, 2001/1/1
Abstract - The observed high-energy cosmic ray proton flux may be used to establish upper limits for the neutrino flux from hadronic sources. In order to obtain the corresponding maximum event rates in next-generation neutrino telescopes, one has to take into account the changes in the neutrino spectrum brought about by neutrino-nucleon interactions in the inner Earth. We show that the integro-differential equations describing these changes can be solved analytically, if one approximates the neutrino interaction cross sections as a sum of delta distributions. In addition, we employ a simple model to

discuss the dependence of the final neutrino spectrum on the strength of neutrino interactions in the high energy regime. After computing the event rates allowed by the various proposed neutrino flux limits, we assess the implications for the possibility of inner Earth tomography by means of high energy neutrinos. Finally, we briefly discuss the significance a non-detection of extragalactic neutrinos in next-generation neutrino telescopes might have.

Implications of Recent Neutrino Oscillation Data
Weiler, T. J.
AIP Conf. Proc. 478: COSMO-98, 1999/1/1, 422
Abstract - Not Available

Implications of the New Neutrino Oscillation Results
Scott, W. G.
The Identification of Dark Matter , 1999/1/1, 540
Abstract - Not Available

Implications of the nu_mu-->nu_s solution to the atmospheric neutrino anomaly for early Universe cosmology
Foot, R.
Astroparticle Physics, 10, 1999/3/1, 253-273
Abstract - By numerically solving the quantum kinetic equations we compute the range of parameters where the nu_mu-->nu_s oscillation solution to the atmospheric neutrino anomaly is consistent with a stringent big bang nucleosynthesis (BBN) bound of N_eff^BBN <~3.6. We show that this requires tau neutrino masses in the range m<INF>nu_tau</INF> >~4<HSP SP = "2">eV (for |deltam^2_atm| = 10^-2.5<HSP SP = "2">eV^2). We discuss the implications of this scenario for hot+cold dark matter, BBN, and the anisotropy of the cosmic microwave background.

Inclusive Neutrino and Antineutrino Reactions in Iron and an Interesting Relation
Mintz, Stephan L.
The Role of Neutrinos, Strings, Gravity,

and Variable Cosmological Constant in Elementary Particle Physics, 2001/1/1, 125
Abstract - Not Available

Influence of Collective Plasma Processes on the Theoretical Flux of Solar Neutrinos
Fedorova, A. V.; Tutukov, A. V.
Astronomy Reports, 46, 2002/3/1, 255-258
Abstract - Not Available

In-Ice Radio Detection of GZK Neutrinos
Seckel, D.
AIP Conf. Proc. 579: Radio Detection of High Energy Particles, 2001/1/1, 196
Abstract - Not Available

Invariant Box-Parameterization of Neutrino Oscillations
Weiler, T. J.; Wagner, D. J.
AIP Conf. Proc. 444: Particle Physics and Cosmology, First Tropical Workshop, 1998/1/1, 46
Abstract - Not Available

Inverse neutrinoless double beta decay and ΔL = 2 processes at linear colliders
Belanger, G.
Lepton and Baryon Number Violation in Particle Physics, neral Astrophysics and Cosmology, 1999/1/1, 605
Abstract - Not Available

IR limits, pregalactic stars, neutrino decay and quantum gravity
Biller, Steve
Astroparticle Physics, 11, 1999/6/1, 103-109
Abstract - The results of recent studies which seek to probe fundamental aspects of astrophysics using TeV gamma-ray observations of active galactic nuclei (AGN) are presented. These analyses take advantage of both spectral and temporal information together with the unique combination of the very high energies and large distance scales involved. The next generation of instruments currently planned are expected to make substantial improvements in the study of such phenomena.

Is neutrino decay really ruled out as a solution to the atmospheric neutrino problem from

Super-Kamiokande data?
Choubey, S.; Goswami, S.
Astroparticle Physics, 14, 2000/9/1, 67-78
Abstract - In this paper, we perform a detailed χ^2-analysis of the 848 days of Super-Kamiokande (SK) atmospheric neutrino data under the assumptions of ν_μ-ν_τ oscillation and neutrino decay. For the latter, we take the most general case of neutrinos with non-zero mixing and consider the possibilities of the unstable component in ν_μ decaying to a state with which it mixes (scenario (a)) and to a sterile state with which it does not mix (scenario (b)). In the first case, Δm^2 (mass squared difference between the two mass states which mix) has to be $/>0.1$ eV^2 from constraints on /K decays, whereas for the second case, Δm^2 can be unconstrained. For case (a), Δm^2 does not enter the χ^2-analysis, while in case (b), it enters the χ^2-analysis as an independent parameter. In scenario (a), there is Δm^2 averaged oscillation in addition to decay, and this gets ruled out at 100.0% CL by the latest SK data. Scenario (b) on the other hand, gives a reasonably good fit to the data for $\Delta m^2 /\sim 0.001$ eV^2.

Is There Time Variation of the Homestake Neutrino Flux?
Lissia, M.
Eighteenth Texas Symposium on Relativistic Astrophysics, 1998/1/1, 706
Abstract - Not Available

KamLAND Experiment-Exploring Low Energy Neutrino Physics
Shirai, J.
The Identification of Dark Matter , 1999/1/1, 590
Abstract - Not Available

Karmen ν_{e} --> ν_{x}
Disappearance Search in a 3 Neutrino Flavor Analysis
Eitel, K.
Particle and Nuclear Physics, 1998/1/1, 203
Abstract - Not Available

Korea Makes a Bid to Catch Neutrinos From the Cosmos
Normile, Dennis
Science, 279, 1998/2/1, 802
Abstract - Not Available

Large neutrino flavour mixings and lepton mass matrices
Tanimoto, M.
Lepton and Baryon Number Violation in Particle Physics, neral Astrophysics and Cosmology, 1999/1/1, 173
Abstract - Not Available

Lepton Asymmetry Generation from Netrino Oscillations in the Early Universe
Semikoz, D. V.
Proceedings of the Sixth SFB-375 Ringberg Workshop Astroteilchenphysik, 2000/2/1, 79
Abstract - Not Available

Limits on Neutrino Masses in SO(10)GUT'S
Pisanti, O.; Rosa, L.
Particle and Nuclear Physics, 1998/1/1, 81
Abstract - Not Available

Long Baseline Neutrino Oscillation Experiments
Zuber, K.
Dark Matter in Astro- and Particle Physics, 2001/1/1, 469
Abstract - Not Available

Looking for Neutralinos From Neutrinos
Chen, Xuelei
News Letter of the Astronomical Society of New York, 5, 1998/8/1, 12
Abstract - Not Available

Low-metal core solar models with helioseismic constraints and the solar neutrino problem
Watanabe, Satoru; Shibahashi, Hiromoto
Proceedings of the SOHO 10/GONG 2000 Workshop: Helio- and asteroseismology at the dawn of the millennium, 2-6 October 2000, Santa Cruz de Tenerife, Tenerife, Spain. Edited by A. Wilson, Scientific coordination by P. L. Pallé. ESA SP-464, Noordwijk: ESA Publications Division, ISBN 92-9092-697-X, 2001, p. 567 - 570, 464, 2001/1/1, 567
Abstract - With the imposition of the

seismically determined sound speed profile, we construct solar models having low metal abundance in the core and evaluate the neutrino fluxes of these models to see if nonstandard solar models with a low metal core can solve the solar neutrino problem. Some of these models are in agreement with the Homestake and the Super-Kamiokande data, but none of these satisfy simultaneously all the neutrino flux data including GALLEX and SAGE and the helioseismically inverted density profile.

LSND Neutrino Oscillation Results and Implications
Louis, W. C.; LSND Collaboration
Particle and Nuclear Physics, 1998/1/1, 151
Abstract - Not Available

MACRO and the Atmospheric Neutrino Problem
Spurio, M.; The MACRO Collaboration
New worlds in astroparticle physics, 2001/1/1, 104
Abstract - Not Available

Many-Valued Relationship Between Solar Flare Activity and Neutrino Flux in GALLEX
Klochek, N.; Nikonova, M.; Sotnikova, R.
ASP Conf. Ser. 206: High Energy Solar Physics Workshop - Anticipating Hess!, 2000/1/1, 88
Abstract - Not Available

Mass signature of supernova nu_{mu} and nu_{tau} neutrinos in SuperKamiokande
Beacom, J. F.; Vogel, P.
Physical Review D, 58, 1998/9/1, 3010
Abstract - The νμ and ντ neutrinos (and their antiparticles) from a Galactic core-collapse supernova can be observed in a water-Cerenkov detector by the neutral-current excitation of ^{16}O. The number of events expected is several times greater than from neutral-current scattering on electrons. The observation of this signal would be a strong test that these neutrinos are produced in core-collapse supernovae, and with the right characteristics. In this paper, this signal is used as the basis for a technique of neutrino mass determination from a future Galactic supernova. The masses of the νμ and ντ neutrinos can either be measured or limited by their delay relative to the $\bar{ν}_e$ neutrinos. By comparing to the high-statistics $\bar{ν}_e$ data instead of the theoretical expectation, much of the model dependence is canceled. Numerical results are presented for a future supernova at 10 kpc as seen in the SuperKamiokande detector. Under reasonable assumptions, and in the presence of the expected counting statistics, νμ and ντ masses down to about 50 eV can be simply and robustly determined. The signal used here is more sensitive to small neutrino masses than the signal based on neutrino-electron scattering.

Mass signature of supernova nu_{mu} and nu_{tau} neutrinos in the Sudbury Neutrino Observatory
Beacom, J. F.; Vogel, P.
Physical Review D, 58, 1998/11/1, 3012
Abstract - Core-collapse supernovae emit of order 10^{58} neutrinos and antineutrinos of all flavors over several seconds, with average energies of 10-25 MeV. In the Sudbury Neutrino Observatory (SNO), which begins operations this year, neutrinos and antineutrinos of all flavors can be detected by reactions which break up the deuteron. For a future Galactic supernova at a distance of 10 kpc, several hundred events will be observed in SNO. The νμ and ν τ neutrinos and antineutrinos are of particular interest, as a test of the supernova mechanism. In addition, it is possible to measure or limit their masses by their delay (determined from neutral-current events) relative to the $\bar{ν}_e$ neutrinos (determined from charged-current events). Numerical results are presented for such a future supernova as seen in SNO. Under reasonable assumptions, and in the presence of the expected counting statistics, a νμ or ν τ mass down to about 30 eV can be simply and robustly determined. If zero delay is measured, then the mass limit is independent of the distance D. At present, this seems to be the best possibility for

direct determination of a ν_μ or ν_τ mass within the cosmologically interesting range. We also show how to separately study the supernova and neutrino physics, and how changes in the assumed supernova parameters would affect the mass sensitivity.

Mass splitting of three seesaw neutrinos
Ma, E.
Weak Interactions and Neutrinos, 2000/1/1, 350
Abstract - Not Available

Massive Neutrinos
Vogel, P.
Neutrino Physics and Astrophysics, 1998/1/1, 213
Abstract - Not Available

Massive Neutrinos as Dark Matter ?? an Overview of Experiments
Wong, H. T.
ASP Conf. Ser. 151: Cosmic Microwave Background and Large Scale Structure of the Universe, 1998/1/1, 129
Abstract - Not Available

Massive Neutrinos in a Coset-Space Family Unification
Yanagida, T.
New Era in Neutrino Physics , 1998/1/1, 65
Abstract - Not Available

Massive neutrinos in physics and astrophysics
Mohapatra, Rabindra N.; Pal, Palash B.
Massive neutrinos in physics and astrophysics, 2nd ed. /Rabindra N. Mohapatra and Palash B. Pal. River Edge, N.J. : World Scientific, c1998. QB 464.2 M63 1999., 1998/1/1
Abstract - Not Available

Massive sterile neutrinos as warm dark matter
Dolgov, A. D.; Hansen, S. H.
Astroparticle Physics, 16, 2002/1/1, 339-344
Abstract - We show that massive sterile neutrinos mixed with the ordinary ones may be produced in the early universe in the right amount to be natural warm dark matter particles. Their mass should be below 40 keV and the corresponding mixing angles $\sin^2 2\theta$ >10^{11} for mixing with ν_μ or ν_τ, while mixing with ν_e is slightly stronger bounded with mass less than 30 keV.

Matter-affected neutrino oscillations in ordinary and mirror stars and their implications for gamma-ray bursts
Volkas, R. R.; Wong, Y. Y. Y.
Astroparticle Physics, 13, 2000/3/1, 21-30
Abstract - It has been proposed that the annihilation process ν<OVL SThe AMANDA neutrino telescope: principle of operation and first results
Andres, E.; Askebjer, P.; Barwick, S. W.; Bay, R.; Bergström, L.; Biron, A.; Booth, J.; Bouchta, A.; Carius, S.; Carlson, M.; Cowen, D.; Dalberg, E.; DeYoung, T.; Ekström, P.; Erlandson, B.; Goobar, A.; Gray, L.; Hallgren, A.; Halzen, F.; Hardtke, R.; Hart, S.; He, Y.; Heukenkamp, H.; Hill, G.; Hulth, P. O.; Hundertmark, S.; Jacobsen, J.; Jones, A.; Kandhadai, V.; Karle, A.; Koci, B.; Lindahl, P.; Liubarsky, I.; Leuthold, M.; Lowder, D. M.; Marciniewski, P.; Mikolajski, T.; Miller, T.; Miocinovic, P.; Mock, P.; Morse, R.; Niessen, P.; Pérez de los Heros, C.; Porrata, R.; Potter, D.; Price, P. B.; Przybylski, G.; Richards, A.; Richter, S.; Romenesko, P.; Rubinstein, H.; Schneider, E.; Schmidt, T.; Schwarz, R.; Solarz, M.; Spiczak, G. M.; Spiering, C.; Streicher, O.; Sun, Q.; Thollander, L.; Thon, T.; Tilav, S.; Walck, C.; Wiebusch, C.; Wischnewski, R.; Woschnagg, K.; Yodh, G.
Astroparticle Physics, 13, 2000/3/1, 1-20
Abstract - AMANDA is a high-energy neutrino telescope presently under construction at the geographical South Pole. In the Antarctic summer 1995/96, an array of 80 optical modules (OMs) arranged on 4 strings (AMANDA-B4) was deployed at depths between 1.5 and 2 km. In this paper we describe the design and performance of the AMANDA-B4 prototype, based on data collected between February and November 1996. Monte Carlo simulations of the detector response to down-going atmospheric muon tracks show that the global behavior of the

detector is understood. We describe the data analysis method and present first results on atmospheric muon reconstruction and separation of neutrino candidates. The AMANDA array was upgraded with 216 OMs on 6 new strings in 1996/97 (AMANDA-B10), and 122 additional OMs on 3 strings in 1997/98.

Measurement of the Solar Electron Neutrino Flux with the Homestake Chlorine Detector
Cleveland, Bruce T.; Daily, Timothy; Davis, Raymond, Jr.; Distel, James R.; Lande, Kenneth; Lee, C. K.; Wildenhain, Paul S.; Ullman, Jack
Astrophysical Journal, 496, 1998/3/1, 505
Abstract - The Homestake Solar Neutrino Detector, based on the inverse beta-decay reaction nu e + 37Cl --> 37Ar + e-, has been measuring the flux of solar neutrinos since 1970. The experiment has operated in a stable manner throughout this time period. All aspects of this detector are reviewed, with particular emphasis on the determination of the extraction and counting efficiencies, the key experimental parameters that are necessary to convert the measured 37Ar count rate to the solar neutrino production rate. A thorough consideration is also given to the systematics of the detector, including the measurement of the extraction and counting efficiencies and the nonsolar production of 37Ar. The combined result of 108 extractions is a solar neutrino-induced 37Ar production rate of 2.56 +/- 0.16 (statistical) +/- 0.16 (systematic) SNU.

Measurements of ^{241}Pu ß-SPECTRUM in Search for Admixture of Massive Neutrinos
Rysavý, M.; Brabec, V.; Dragoun, O.; Dragounová, N.; Spalek, A.; Ízek, J.; Kovalík, A.; Novgorodov, A. F.
Particle and Nuclear Physics, 1998/1/1, 335
Abstract - Not Available

MINOS: a long baseline neutrino oscillation experiment at NUMI
Para, A.
Weak Interactions and Neutrinos,

2000/1/1, 410
Abstract - Not Available

Model Independent Analysis of the Solar Neutrino Data
Minakata, H.; Nunokawa, H.
New Era in Neutrino Physics , 1998/1/1, 209
Abstract - Not Available

Modification of Neutrino Reaction Rates in Hot Dense Matter
Yamada, S.
LNP Vol. 287: Proceedings of the 9th workshop on Nuclear Astrophysics, 1998/1/1, 115
Abstract - Not Available

MSW Time Variations of the Solar Neutrino flux
Malyshkin, L. M.; Kulsrud, R. M.
American Astronomical Society Meeting, 193, 1998/12/1
Abstract - We consider the possibility of MSW time variations of the solar neutrino flux due to neutrino interference effects. For two types of neutrinos there are two MSW neutrino solutions, + and -. In vacuum these are the m_1 and m_2 (m_1<m_2) mass neutrino states. These two neutrino states have propagation time delays relative to light, t_+ and t_-, due to nonzero rest masses. Let sigma_t equiv hbar /|t_+-t_-|. The energy difference of the two states is sigma_a ={cal E} a|t_+-t_-|/c, where {cal E} is the emitted neutrino energy, and a is the acceleration of the emitting nucleus. Two necessary criteria to observe neutrino flux time variations are proposed. The first criterion, \sigma < sigma_t, for the time variation not to be averaged to zero over the detection energy width sigma , should be satisfied. In the case of the 0.862 Mev (7) Be solar neutrino line this criterion gives us an upper estimate for m_2(2-m_1^2) , the difference of neutrino mass squared, of ~ 10(-9) ev(2) . The second criterion, \sigma_a < \sigma_t, is necessary for collisional coherence to exist. This criterion is equivalent to a plausible condition a|t_+(2-t_-^2|) < lambda_. Our criterion is stronger than collisional coherence criteria

formerly suggested, which are roughly equivalent to the condition $a|t_+{}^{\wedge}2 - t_-{}^{\wedge}2|$ < 1, because the neutrino De Broglie wavelength $lambda_$ is much less than the interparticle distance in the solar interior l. Exact values for the detected neutrino flux and its variations will be presented for both the case of a solar neutrino line, and the case of a continuous neutrino spectrum with a Gaussian shape of the energy response function of the neutrino detection device.

Mu- and Tau-Neutrino Spectra Formation in Supernovae
Raffelt, Georg G.
Astrophysical Journal, 561, 2001/11/1, 890-914
Abstract - The μ- and τ-neutrinos emitted from a proto-neutron star are produced by nucleonic bremsstrahlung NN-->NNνν&dl; and pair annihilation e^{+e-}-->νν&dl; reactions that freeze out at the "energy sphere." Before escaping from there to infinity, the neutrinos diffuse through the "scattering atmosphere," a layer in which their main interaction is elastic scattering on nucleons νN-->Nν. If these collisions are taken to be isoenergetic, as in all numerical supernova simulations, the neutrino flux spectrum escaping to infinity depends only on the medium temperature T_{ES} and the thermally averaged optical depth τ&dl;$_{ES}$ at the energy sphere.

Multi-dimensional Simulations of Core Collapse Supernovae employing Ray-by-Ray Neutrino Transport
Hix, W. R.; Mezzacappa, A.; Liebendoerfer, M.; Messer, O. E. B.; Blondin, J. M.; Bruenn, S. W.
American Astronomical Society Meeting, 199, 2001/12/1
Abstract - Decades of research on the mechanism which causes core collapse supernovae has evolved a paradigm wherein the shock that results from the formation of the proto-neutron star stalls, failing to produce an explosion. Only when the shock is re-energized by the tremendous neutrino flux that is carrying off the binding energy of this proto-neutron star can it drive off the star's envelope, creating a supernova. Work in recent years has demonstrated the importance of multi-dimensional hydrodynamic effects like convection to successful simulation of an explosion. Further work has established the necessity of accurately characterizing the distribution of neutrinos in energy and direction. This requires discretizing the neutrino distribution into multiple groups, adding greatly to the computational cost. However, no supernova simulations to date have combined self-consistent multi-group neutrino transport with multi-dimensional hydrodynamics. We present preliminary results of our efforts to combine these important facets of the supernova mechanism by coupling self-consistent ray-by-ray multi-group Boltzmann and flux-limited diffusion neutrino transport schemes to multi-dimensional hydrodynamics. This research is supported by NASA under contract NAG5-8405, by the NSF under contract AST-9877130, and under a SciDAC grant from the DoE Office of Science High Energy and Nuclear Physics Program. Work at Oak Ridge National Laboratory is managed by UT-Battelle, LLC, for the U.S. Department of Energy under contract DE-AC05-00OR22725.

Multi-GEV Neutrinos from Internal Dissipation in Gamma-Ray Burst Fireballs
Mészáros, P.; Rees, M. J.
Astrophysical Journal, 541, 2000/9/1, L5-L8
Abstract - Subphotospheric internal shocks and transverse differences of the bulk Lorentz factor in relativistic fireball models of gamma-ray bursts (GRBs) lead to neutron diffusion relative to protons, resulting in inelastic nuclear collisions. This results in significant fluxes of ν$_{>μ}$ (ν&dl; $_{μ}$) of ~3 GeV and ν$_e$ (ν&dl; $_e$) of ~2 GeV, scaling with the flow Lorentz factor η<η$_{π}$~400. This extends significantly the parameter space for which neutrinos from inelastic collision are expected, which in the absence of the above effects requires values in excess of η$_{>π}$. A model with sideways diffusion of neutrons from a

slower wind into a fast jet can lead to production of ν $_{μ}$(ν&dl;$_{μ}$) and ν$_e$(ν&dl;$_e$) in the range 2-25 GeV or higher, depending on the value of η. The emission from either of these mechanisms from GRBs at redshifts $z\sim1$ may be detectable in suitably densely spaced detectors.

Muon Measurements in the Atmosphere in the Context of the Atmospheric Neutrino Anomaly
Circella, M.
New Vistas in Astrophysics, 2000/1/1, 85
Abstract - Not Available

Muon Neutrinos with the MACRO Detector at L.N.G.S.
Montaruli, T.; The MACRO Collaboration
Particle and Nuclear Physics, 1998/1/1, 249
Abstract - Not Available

Nearly degenerate neutrino masses and nearly decoupled neutrino oscillations
Fritzsch, H.; Xing, Zhi-Zhong
Weak Interactions and Neutrinos, 2000/1/1, 232
Abstract - Not Available

Neutral-Current Detection via ^{3}He (n.p)^{3}H in the Sudbury Neutrino Observatory
Robertson, R. G. H.
Particle and Nuclear Physics, 1998/1/1, 113
Abstract - Not Available

Neutrino (Mu) --> Neutrino(S) Oscillations of Atmospheric Neutrinos and Neutrino Mass Spectrum
Liu, Q.-Y.; Smirnov, A. Yu.
New Era in Neutrino Physics , 1998/1/1, 229
Abstract - Not Available

Neutrino absorption tomography of the Earth's interior using isotropic ultra-high energy flux
Jain, Pankaj; Ralston, John P.; Frichter, George M.
Astroparticle Physics, 12, 1999/11/1, 193-198
Abstract - We study the feasibility of using an isotropic flux of cosmic neutrinos in the energy range of 10 to 10000 TeV to study the interior structure of Earth. The angular distribution of events in a ~km^3-scale neutrino telescope can be inverted to yield information on the Earth's mass distribution that is independent of other methods. The energy spectrum of the neutrino primaries is also determined from consistency with the angular distribution. It is possible to make a model independent determination of the density profile of Earth's interior, separate from the absolute normalization of the incident cosmic neutrinos.

Neutrino Afterglow from Gamma-Ray Bursts: $\sim10^{18}$ EV
Waxman, Eli; Bahcall, John N.
Astrophysical Journal, 541, 2000/10/1, 707-711
Abstract - We show that a significant fraction of the energy of a γ-ray burst (GRB) is probably converted to a burst of $10^{17\text{-}10/19}$ eV neutrinos and multiple GeV γ-rays that follow the main GRB by ~10 s. If GRBs accelerate protons to $\sim10^{20}$ eV, a suggestion that recently gained support from observations of GRB afterglows, then both the neutrinos and the γ-rays may be detectable.

Neutrino Afterglows and Progenitors of Gamma-Ray Bursts
Dai, Z. G.; Lu, T.
Astrophysical Journal, 551, 2001/4/1, 249-253
Abstract - Currently popular models for progenitors of gamma-ray bursts (GRBs) are the mergers of compact objects and the explosions of massive stars. These two cases have distinctive environments for GRBs: compact object mergers occur in the interstellar medium (ISM) and the explosions of massive stars occur in the preburst stellar wind. We here discuss neutrino afterglows from reverse shocks as a result of the interaction of relativistic fireballs with their surrounding wind matter. After comparing with the analytical result of Waxman & Bahcall for the homogeneous ISM case, we find that the

differential spectrum of neutrinos with energy from $\sim 3 \times 10^{15}$ to $\sim 3 \times 10^{17}$ eV in the wind case is softer by 1 power of the energy than in the ISM case. Furthermore, the expected flux of upward moving muons produced by neutrino interactions below a detector on the surface of the Earth in the wind case is ~ 5 events yr^{-1}km^{-2}, which is about 1 order of magnitude larger than in the ISM case. In addition, these properties are independent of whether the fireballs are isotropic or beamed. Therefore, neutrino afterglows, if detected, may provide a way of distinguishing between GRB progenitor models based on the differential spectra of neutrinos and their event rates in a detector.

Neutrino Annihilation between Binary Neutron Stars
Salmonson, Jay D.; Wilson, James R.
Astrophysical Journal, 561, 2001/11/1, 950-956
Abstract - We calculate the neutrino pair annihilation rate into electron pairs between two neutron stars in a binary system. We present a closed formula for the energy deposition rate at any point between the stars, where each neutrino of a pair derives from each star, and compare this result with that in which all neutrinos derive from a single neutron star. An approximate generalization of this formula is given to include the relativistic effects of gravity. We find that this interstar neutrino annihilation is a significant contributor to the energy deposition between heated neutron star binaries. In particular, for two neutron stars near their last stable orbit, interstar neutrino annihilation energy deposition is almost equal to that of single-star energy deposition.

Neutrino Astronomy and Indirect Search for WIMPs
Montaruli, T.
Dark Matter in Astro- and Particle Physics, 2001/1/1, 688
Abstract - Not Available

Neutrino Astronomy and the Amanda South Pole Telescope
Halzen, Francis
New Vistas in Astrophysics, 2000/1/1, 3
Abstract - Not Available

Neutrino Astronomy in Experiment and Theory
Rhode, W.
Astronomische Gesellschaft Meeting Abstracts, 15, 1999/1/1
Abstract - This contribution presents an overview of the present status of high energy neutrino astrophysics with special emphasis on the connection to the classical astronomy. Included are descriptions of the possible neutrino sources, acceleration processes, methods of detection, and existing and planned experiments. Recent results of the existing experiments are discussed.

Neutrino Astronomy with the MACRO Detector
Ambrosio, M.; Antolini, R.; Auriemma, G.; Bakari, D.; Baldini, A.; Barbarino, G. C.; Barish, B. C.; Battistoni, G.; Bellotti, R.; Bemporad, C.; Bernardini, P.; Bilokon, H.; Bisi, V.; Bloise, C.; Bower, C.; Brigida, M.; Bussino, S.; Cafagna, F.; Calicchio, M.; Campana, D.; Carboni, M.; Cecchini, S.; Cei, F.; Chiarella, V.; Choudhary, B. C.; Coutu, S.; De Cataldo, G.; Dekhissi, H.; De Marzo, C.; De Mitri, I.; Derkaoui, J.; De Vincenzi, M.; di Credico, A.; Enriquez, O.; Favuzzi, C.; Forti, C.; Fusco, P.; Giacomelli, G.; Giannini, G.; Giglietto, N.; Giorgini, M.; Grassi, M.; Gray, L.; Grillo, A.; Guarino, F.; Gustavino, C.; Habig, A.; Hanson, K.; Heinz, R.; Iarocci, E.; Katsavounidis, E.; Katsavounidis, I.; Kearns, E.; Kim, H.; Kyriazopoulou, S.; Lamanna, E.; Lane, C.; Levin, D. S.; Lipari, P.; Longley, N. P.; Longo, M. J.; Loparco, F.; Maaroufi, F.; Mancarella, G.; Mandrioli, G.; Manzoor, S.; Margiotta, A.; Marini, A.; Martello, D.; Marzari-Chiesa, A.; Mazziotta, M. N.; Michael, D. G.; Mikheyev, S.; Miller, L.; Monacelli, P.; Montaruli, T.; Monteno, M.; Mufson, S.; Musser, J.; Nicolò, D.; Nolty, R.; Okada, C.; Orth, C.; Osteria, G.; Ouchrif, M.; Palamara, O.; Patera, V.; Patrizii, L.; Pazzi, R.; Peck, C. W.; Perrone, L.; Petrera, S.; Pistilli, P.; Popa, V.; Rainò, A.; Reynoldson, J.; Ronga, F.; Satriano, C.; Satta, L.; Scapparone, E.;

Scholberg, K.; Sciubba, A.; Serra, P.; Sioli, M.; Sitta, M.; Spinelli, P.; Spinetti, M.; Spurio, M.; Steinberg, R.; Stone, J. L.; Sulak, L. R.; Surdo, A.; Tarlè, G.; Togo, V.; Vakili, M.; Vilela, E.; Walter, C. W.; Webb, R.
Astrophysical Journal, 546, 2001/1/1, 1038-1054
Abstract - High-energy gamma-ray astronomy is now a well-established field, and several sources have been discovered in the region from a few giga-electron volts up to several tera-electron volts. If sources involving hadronic processes exist, the production of photons would be accompanied by neutrinos too. Other possible neutrino sources could be related to the annihilation of weakly interacting, massive particles (WIMPs) at the center of galaxies with black holes. We present the results of a search for pointlike sources using 1100 upward-going muons produced by neutrino interactions in the rock below and inside the Monopole Astrophysics and Cosmic Ray Observatory (MACRO) detector in the underground Gran Sasso Laboratory. These data show no evidence of a possible neutrino pointlike source or of possible correlations between gamma-ray bursts and neutrinos. They have been used to set flux upper limits for candidate pointlike sources which are in the range 10^{-14}-10^{-15} cm^{-2} s^{-1}.

Neutrino Astronomy: First Light--Using The South Pole IceCap
Morse, R. M.
American Astronomical Society Meeting, 192, 1998/5/1
Abstract - Astronomy is now able to investigate the photon spectrum from radio frequencies (the Cosmic Microwave Background Radiation--CMBR) to the highest-energy gamma rays using ground-based gamma-ray telescopes. However, because of the intergalactic CMBR and infra-red light (IR), space becomes increasing opaque to these high-energy photons as their energy increases. Neutrinos have a distinct advantage in this energy regime, for unlike photons of energies of 10(13) eV and beyond, they are not absorbed by the intergalactic IR light

and CMBR. Neutrinos can, in principle, reach us from the edge of the universe, whereas, if the sources of the high-energy cosmic rays were beyond the Virgo cluster, then the conventional window of exploration would be closed above 100 TeV. Thus, sources which may be responsible for the highest-energy cosmic rays may lie beyond the discovery radius of conventional astronomy.

Neutrino Astronomy: First Light-Using The South Pole IceCap
Morse, R. M.
ASP Conf. Ser. 141: Astrophysics From Antarctica, 1998/1/1, 10
Abstract - Not Available

Neutrino Astronomy: The Case for Antares
Moscoso, L.
AIP Conf. Proc. 415: Beyond the Standard Model. From Theory to Experiment, 1998/1/1, 292
Abstract - Not Available

Neutrino astrophysics and cosmology
Abazajian, Kevork Nazar Ph.D.
Thesis, 2001/1/1, 1
Abstract - Although physical cosmology is becoming a field rich in data, the theoretical basis for several aspects of standard cosmological models are spectacularly devoid of firm foundations. On the other hand, the standard model of particle physics has successfully described an enormous quantity of experimental data, with one exception lying in the neutrino sector from observations of the atmospheric neutrino flux. This dissertation intersects both fields, as an interplay of the problems confronting theoretical cosmology and the tremendous success of the standard model of particle physics. And, in return, the successes of the standard cosmology may give insights into new particle physics, particularly neutrino physics. In this interplay, this dissertation studies the production of sterile neutrino dark matter in the early universe, constraints on this scenario, including radiative decays in galactic clusters. The effects of nonthermal neutrinos resulting from neutrino

transformation on big bang nucleosynthesis are studied. Also investigated are the possible constraints on neutrino mixing parameters from neutrino mixing in the early universe. And, the effects on big bang nucleosynthesis of baryon and antibaryon inhomogeneities resulting from some models of baryon creation (baryogenesis) are studied.

Neutrino Astrophysics at the Cross Roads
Raffelt, G. G.
High Energy Physics and Cosmology, 1998 Summer School, 1999/1/1, 147
Abstract - Not Available

Neutrino Astrophysics with the MACRO Detector
Montaruli, Teresa; Cei, Fabrizio; Pazzi, Roberto; Ronga, Francesco; The MACRO Collaboration
Gamma-ray Bursts in the Afterglow Era, 2001/1/1, 75
Abstract - A sample of 1197 neutrino induced upward-going muons is used to look for point-like sources and for space-time correlations with BATSE and BeppoSAX gamma-ray bursts. We set an upper limit (90% c.l.) of 6.9 times 10^{-10} cm^{-2} upward-going muons per average burst. A search for high multiplicity events in coincidence with GRBs has been performed using the supernova trigger.

Neutrino Breakout Bursts in Stellar Collapse
Burrows, A.
American Astronomical Society Meeting, 200, 2002/5/1
Abstract - I will present some of the results of a series of simulations of stellar collapse with a multi-group, multi-angle Feautrier transport code, using the tangent-ray method to derive the angular dependence of the radiation field. The spectral, species, and temporal character of the burst and the dependence of its features on progenitor mass and physics inputs will be reviewed, as will be the detectability of the burst by various underground neutrino telescopes. Support for this work was provided by the Scientific Discovery through Advanced Computing (SciDAC) program of the

DOE, grant number DE-FC02-01ER41184.

Neutrino Burst from Supernovae and Neutrino Oscillation
Sato, K.
Eighteenth Texas Symposium on Relativistic Astrophysics, 1998/1/1, 720
Abstract - Not Available

Neutrino Conversion and Neutrino Astrophysics
Smirnov, A. Yu.
New Era in Neutrino Physics , 1998/1/1, 1
Abstract - Not Available

Neutrino conversions in cosmological gamma-ray burst fireballs
Athar, H.
Astroparticle Physics, 14, 2000/11/1, 217-225
Abstract - We study neutrino conversions in a recently envisaged source of high energy neutrinos (E>~10_6 GeV), i.e., in the vicinity of cosmological gamma-ray burst fireballs (GRB). We consider the effects of flavor and spin-flavor conversions and point out that, in both situations, a some what higher than estimated high energy tau neutrino flux from GRBs is expected in new km^2 surface area under water/ice neutrino telescopes.

Neutrino Conversions in Solar Random Magnetic Fields
Torrente-Lujan, E.
The Chaotic Universe, Proceedings of the Second ICRA Network Workshop, Advanced Series in Astrophysics and Cosmology, vol.10, Edited by V. G. Gurzadyan and R. Ruffini, World Scientific, 2000, p.646, 2000/1/1, 646
Abstract - Not Available

Neutrino Cooling of Neutron Stars: Medium Effects
Voskresensky, Dmitri N.
LNP Vol. 578: Physics of Neutron Star Interiors, 2001/1/1, 467
Abstract - This review demonstrates that the neutrino emission from the dense hadronic component in neutron stars is subject to strong modifications due to collective effects in nuclear matter. With

the most important in-medium processes incorporated in the cooling code an overall agreement with available soft X ray data can be easily achieved. With these findings so called "standard" and "on-standard" cooling scenarios are replaced by one general "nuclear medium cooling scenario" which relates slow and rapid neutron star coolings to the star masses (interior densities). In-medium effects play an important role also in the early hot stage of the neutron star evolution by decreasing the neutrino opacity for less massive and increasing it for more massive neutron stars. A formalism for the calculation of the neutrino radiation from nuclear matter is presented that treats on equal footing one-nucleon and multiple-nucleon processes as well as reactions with resonance bosons and condensates. The cooling history of neutron stars with quark cores is also discussed.

Neutrino Cosmology
Dolgov, A. D.
NATO ASIC Proc. 511: Current Topics in Astrofundamental Physics: Primordial Cosmology, 1998/1/1, 685
Abstract - Not Available

Neutrino Dark Matter
Caldwell, D. O.
New Worlds in Astroparticle Physics II, 1999/1/1, 257
Abstract - Not Available

Neutrino dark matter search at accelerators
Vander Donckt, M.
Dark matter in Astrophysics and Particle Physics, 1999/1/1, 830
Abstract - Not Available

Neutrino decay and the thermochemical equilibrium of the interstellar medium
Sánchez, N. M.; Añez, N. Y.
Astronomy and Astrophysics, 354, 2000/2/1, 1123-1126
Abstract - We calculate the thermochemical equilibrium of the diffuse interstellar medium, including ionization by a photon flux F_{nu} from neutrino decay. The main heating mechanism considered is photoelectrons from grains and PAHs. For the studied range of F_{nu} values, there always exists two regions of stability (a warm and a cold phase) that can coexist in equilibrium if the thermal interstellar pressure is between a maximum value (P_{max}) and a minimum value (P_{min}). High F_{nu} values ($\sim 10^4$-10^5 $cm^{-2}s^{-1}$) can be consistent with observed interstellar pressures only if more efficient sources are heating the gas. It is shown that a neutrino flux increase (due, for example, to an increase in the supernova explosion rate) may stimulate the condensation of cold gas by decreasing P_{max} below the interstellar pressure value.

Neutrino deficit challenges conservation laws
Wilczek, Frank
Nature, 391, 1998/1/1, 123
Abstract - How an anomaly in the relative number of electron-type and muon-type neutrinos found in cosmic rays threatens the law of lepton number conservation is discussed. The anomaly calls into question the completeness of the Standard Model of particle physics.

Neutrino Degeneracy and Chemically Inhomogeneous Universe
Dolgov, A. D.
AIP Conf. Proc. 478: COSMO-98, 1999/1/1, 259
Abstract - Not Available

Neutrino Degeneracy in Big-Bang Nucleosynthesis and Cosmic Rays: Evolution of the Light Elements
Kajino, T.; Orito, M.; Tokuhisa, A.
International Symposium on Origin of Matter and Evolution of Galaxies 97, 1998/1/1, 27
Abstract - Not Available

Neutrino Detectives: Digging Deep at the Sudbury Neutrino Observatory
Kulyk, Christine
Journal of the Royal Astronomical Society of Canada, 95, 2001/10/1, 185
Abstract - Not Available

Neutrino Electron Plasma Instability
Lai, C. H.; Tajima, T.
American Astronomical Society Meeting,

195, 1999/12/1
Abstract - Weak interactions play an important role in early universe plasma collection, especially on neutrinos and the corresponding leptons. It also has important effects on the detail balance of the primordial nucleosynthesis, especially on the production of He and light elements. At around temperature T=300 GeV, the primordial plasma undergone electroweak spontaneous symmetry breaking (SSB) phase transition. Some of the gauge bosons and other particles gain mass via Higgs mechanism. Deduced from Weinberg-Salam electroweak theory, a Boltzmann equation and subsequent fluid equations are derived for the primordial electron-positron-neutrino-photon plasma. A collective instability that separates the phases of electrons(and positrons) and neutrinos(and anti-neutrinos) is discussed. We also discussed the application of the fluctuation-dissipation theory in this system of plasma. Astrophysical applications and implications are explored, particularly in supernovae and possibly in GRB(Gamma Ray Bursts).

Neutrino electron plasma instability
Lai, Chi-Hsuan Ph.D.
Thesis, 1999/1/1, 16
Abstract - Weak interactions play an important role in early universe plasma collection, especially on neutrinos and the corresponding leptons. It also has important effects on the detail balance of the primordial nucleosynthesis, especially on the production of He and light elements. At around T = 300 GeV, the primordial plasma undergone electroweak spontaneous symmetry breaking (SSB) phase transition. Some of the gauge bosons and other particles gain mass via Higgs mechanism. Deduced from Weinberg-Salam electroweak theory, a Boltzmann equation and subsequent fluid equations are derived for the primordial electron-positron-neutrino- photon plasma. A collective instability that separates the phases of electrons (and positrons) and neutrinos (and anti-neutrinos) is discussed. We also discussed the application of the fluctuation-dissipation theory in this system of plasma. An approach with Hubble expansion included in Boltzmann equation is also discussed. Astrophysical applications and implications are explored, particularly in supernovae.

Neutrino electron scattering and electroweak gauge structure: probing the masses of a new Z boson
Miranda, O. G.; Semikoz, V. B.; Valle, J. W. F.
Lepton and Baryon Number Violation in Particle Physics, neral Astrophysics and Cosmology, 1999/1/1, 683
Abstract - Not Available

Neutrino emission due to Cooper pairing of nucleons in cooling neutron stars
Yakovlev, D. G.; Kaminker, A. D.; Levenfish, K. P.
Astronomy and Astrophysics, 343, 1999/3/1, 650-660
Abstract - The neutrino energy emission rate due to formation of Cooper pairs of neutrons and protons in the superfluid cores of neutron stars is studied. The cases of singlet-state pairing with isotropic superfluid gap and triplet-state pairing with anisotropic gap are analysed. The neutrino emission due to the singlet-state pairing of protons is found to be greatly suppressed with respect to the cases of singlet- or triplet-state pairings of neutrons. The neutrino emission due to pairing of neutrons is shown to be very important in the superfluid neutron-star cores with the standard neutrino luminosity and with the luminosity enhanced by the direct Urca process. It can greatly accelerate both, standard and enhanced, cooling of neutron stars with superfluid cores. This enables one to interpret the data on surface temperatures of six neutron stars, obtained by fitting the observed spectra with the hydrogen atmosphere models, by the standard cooling with moderate nucleon superfluidity.

Neutrino Emission due to Cooper Pairing of Nucleons in Neutron Stars
Yakovlev, D. G.; Kaminker, A. D.; Levenfish, K. P.
Neutron Stars and Pulsars: Thirty Years

after the Discovery, 1998/1/1, 195
Abstract - Not Available

Neutrino emission due to proton pairing in
neutron stars
Kaminker, A. D.; Haensel, P.; Yakovlev,
D. G.
Astronomy and Astrophysics, 345,
1999/5/1, L14-L16
Abstract - We calculate the neutrino energy
emission rate due to singlet-state pairing of
protons in the neutron star cores taking into
account the relativistic correction to the
non-relativistic rate. The non-relativistic
rate is numerically small, and the
relativistic correction appears to be about
10-50 times larger. It plays thus the leading
role, reducing great difference between the
neutrino emissions due to pairing of
protons and neutrons. The results are
important for simulations of neutron star
cooling.

Neutrino emission from neutron stars
Yakovlev, D. G.; Kaminker, A. D.;
Gnedin, O. Y.; Haensel, P.
Physics Reports, 354, 2001/11/1, 1-2
Abstract - Electronic Article Available
from Elsevier Science.

Neutrino Event Rates from Gamma-Ray Bursts
Halzen, F.; Hooper, D. W.
Astrophysical Journal, 527, 1999/12/1,
L93-L96
Abstract - We recalculate the diffuse flux
of high-energy neutrinos produced by
gamma-ray bursts in the relativistic fireball
model. Although we confirm that the
average single burst produces only
~10⁻² high-energy neutrino
events in a detector with a 1
km² effective area, i.e.,
about 10 events yr⁻¹, we
show that the observed rate is dominated
by burst-to-burst fluctuations that are very
large. We find event rates that are expected
to be larger by 1 order of magnitude, likely
more, which are dominated by a few very
bright bursts. This greatly simplifies their
detection.

Neutrino Factory Detector and Long Baseline
Oscillations

Fenyves, E. J.; Burkart, R. F.
20th Texas Symposium on relativistic
astrophysics, 2001/1/1, 873
Abstract - Not Available

Neutrino Flavor Mixing and Oscillations in
Field Theory
Sassaroli, E.
Particles, Strings and Cosmology
(PASCOS 98), 1999/1/1, 310
Abstract - Not Available

Neutrino flavor transformation in supernovae
and the early universe
Fuller, George M.
Current aspects of neutrino physics,
2001/1/1, 255
Abstract - Contents: 1. Introduction. 2.
Matter-enhanced neutrino conversion in
nonlinear environments. 3. Core collapse
supernovae. 4. The early universe and
cosmology.

Neutrino Fluence after r-Process Freezeout and
Abundances of TE Isotopes in Presolar
Diamonds
Qian, Y.-Z.; Vogel, P.; Wasserburg, G. J.
Astrophysical Journal, 513, 1999/3/1, 956-
960
Abstract - Using the data of Richter, Ott,
& Begemann on Te isotopes in
diamond grains from a meteorite, we
derive bounds on the neutrino fluence and
the decay timescale of the neutrino flux
relevant for the supernova r-process. Our
new bound on the neutrino fluence ℱ
after freezeout of the r-process peak at
mass number A~130 is more stringent than
the previous bound ℱ<~0.045 (in
units of 10^37 ergs cm^-2) of Qian et al.
and Haxton et al., if the neutrino flux
decays on a timescale tau&d4;>~0.65 s.
In particular, it requires that a fluence of
ℱ=0.031 be provided by a neutrino
flux with tau&d4;<~0.84 s. Such a
fluence may be responsible for the
production of the solar r-process
abundances at A=124-126. Our results are
based on the assumption of Ott that only
the stable nuclei implanted into the
diamonds are retained, while the
radioactive nuclei are lost from the
diamonds upon decay after implantation.

The mechanisms of preferential retention/loss of the implanted nuclei are not well understood.

Neutrino flux from observable Gamma Ray Bursts
Spada, M.; Guetta, D.; Waxman, E.
AAS/High Energy Astrophysics Division, 32, 2000/12/1
Abstract - We derive the flux and spectrum of neutrinos from Gamma Ray Bursts (GRBs), and the corresponding detection rate in a cubic-km neutrino detector, within the frame work of the Internal Shock Model. In this model, GRBs are produced by internal shocks in a highly relativistic wind, and high energy neutrinos result from photo-meson interactions of wind protons with gamma-ray photons. We show that the predicted neutrino flux is only weakly dependent on unknown wind parameters, due to the fact that observed GRB characteristics require these parameters to be strongly correlated. Thus, the predicted neutrino luminosity does not vary strongly from burst to burst. Several tens of events per year, correlated with GRBs, are expected to be detected in a cubic-km detector.

Neutrino Heating in an Inhomogeneous Big Bang Nucleosynthesis Model
Lara, J. F.
20th Texas Symposium on relativistic astrophysics, 2001/1/1, 349
Abstract - Not Available

Neutrino Induced Waves in Degenerate Electron Plasmas
Laming, J. Martin
AAS/High Energy Astrophysics Division, 31, 1999/4/1
Abstract - Supernovae emit far more energy as neutrinos (10(53) - 10(54) erg) than as electromagnetic radiation ($\sim$ 10(51) erg). A recent paper (Bingham et al 1994) speculated that the intense flux of neutrinos generated during a supernova explosion may interact collisionlessly with the surrounding plasma providing an extra heat source through the excitation of plasma waves. This result was criticised on various grounds by Hardy & Melrose (1997). I will present a discussion of these works and suggest that under certain conditions, neutrino induced waves may possibly be generated with growth rates large enough to be interesting during the few second duration of a supernova neutrino burst. The essential requirements are that the plasma electrons be degenerate (in order to avoid the collisional wave damping that would otherwise result) and that a magnetic field of at least $\sim$ 10(15) G exist. This magnetic field is similar to those recently inferred from observations of magnetars (Kouveliotou et al. 1998, Vasisht & Gotthelf 1997), and so can be considered as within the realm of plausibility.

Neutrino induced waves in degenerate electron plasmas: a mechanism in supernovae or gamma ray bursts?
Laming, J. Martin
New Astronomy, 4, 1999/10/1, 389-403
Abstract - We consider a process related to the inverse of photon/plasmon decay to neutrino-antineutrino pairs; the stimulated emission of photons and plasmons by the intense flux of neutrinos and antineutrinos present during the neutrino burst of a core collapse supernova. We find physically insignificant emission and plasma heating unless a magnetic field is present, with strength >10^15 G. For plasma electron temperatures $\sim$ 1 MeV, a population of waves enhanced over thermal levels in the postshock plasma is also required. The magnetic field is consistent with recent theory and observations concerning highly magnetic pulsars, known as magnetars. The status of this field is briefly reviewed, and the possibility that this instability might also apply to models of gamma ray bursts is also explored.

Neutrino Kinetics in Dense Astrophysical Plasmas
Silva, L. O.; Bingham, R.; Dawson, J. M.; Mendonça, J. T.; Shukla, P. K.
Astrophysical Journal Supplement Series, 127, 2000/4/1, 481-484
Abstract - Starting from the effective potentials describing the interaction of a single neutrino with a background species

and the concept of ponderomotive force of neutrinos, we develop a collisionless relativistic kinetic theory for neutrinos coupled with a background medium. This formalism provides the basic tool for the physics of collective neutrino-plasma interactions. Comparison with equivalent kinetic equations derived using the methods of finite-temperature quantum field theory is presented. Astrophysical scenarios in which collective neutrino-plasma interactions can play an important role are discussed.

Neutrino Magnetic Moment and Supernovae
Ayala, A.; D'Olivo, J. C.; Torres, M.
Particles and Fields, Eighth Mexican School, 1999/1/1, 319
Abstract - Not Available

Neutrino magnetic moments and atmospheric neutrinos
Kim, J. E.
Weak Interactions and Neutrinos, 2000/1/1, 263
Abstract - Not Available

Neutrino Mass
Nagashima, Yorikiyo
Probing Luminous and Dark Matter, 2000/1/1, 174
Abstract - Not Available

Neutrino Mass and Dark Matter
Caldwell, D. O.
AIP Conf. Proc. 444: Particle Physics and Cosmology, First Tropical Workshop, 1998/1/1, 82
Abstract - Not Available

Neutrino Mass and Dark Matter
Caldwell, D. O.
The Identification of Dark Matter , 1999/1/1, 527
Abstract - Not Available

Neutrino mass and its implications for the zero mode and vacuum structures of the standard model and its extensions
Stojkovic, Dejan Ph.D.
Thesis, 2001/1/1, 26
Abstract - In this thesis we explore different field theory models containing massive neutrinos. We analyze the theoretical background of neutrino masses and mixings and consequences for the zero mode and vacuum structures of the standard model and its various extensions. We first consider the most general neutrino masses and mixings including Dirac mass terms, M_D, as well as right- and left-handed Majorana masses, M_R and M_L. We show that for three generations, diagonalization of the Hamiltonian to obtain the propagating eigenstates in the general case requires diagonalization of a 12 x 12 Hermitian matrix, rather than the traditional 6 x 6 symmetric mass matrix. Although the standard "see-saw" mechanism remains valid, and the eigenvalues obtained are identical to the standard ones, the correct description of diagonalization and mixing is more complicated. The analogs of the CKM matrix for the light and the heavy neutrinos need not to be unitary, enriching the opportunities for CP violation in the full neutrino sector. A zero energy solution for Standard Model massless neutrinos on electroweak Z-string has been found, but was thought to be non-normalizable. Here we show that although this mode is not discretely normalizable, it is delta-function normalizable and the correct interpretation of this solution is within the framework of the continuum spectrum.

Neutrino mass and oscillations
Kajita, T.
Space Science Reviews, 100, 2002/1/1, 221-233
Abstract - Present understanding of neutrino mass and mixing are discussed based on recent neutrino experiments, including solar, atmospheric, reactor and long baseline accelerator neutrino experiments. Especially, the results from the Super-Kamiokande experiment are discussed in detail.

Neutrino Mass from Tritium β-Decay
Weinheimer, C.
Dark Matter in Astro- and Particle Physics, 2001/1/1, 513
Abstract - Not Available

Neutrino mass varying with time
Barshay, Saul; Kreyerhoff, Georg
Astroparticle Physics, 10, 1999/1/1, 107-113
Abstract - Experiments and astronomical observations may eventually show that there are comparable energy densities in the present universe from neutrino mass and from an effective cosmological constant, each at the level of a few per cent of the closure density, and similar to the empirical energy density in baryons. We consider the possibility that the cause of such a coincidence is related to a spontaneous breakdown of CP invariance characterized by a cosmological energy scale. Such a spontaneous breakdown can give rise to a neutrino-antineutrino asymmetry just prior to the time of electroweak symmetry breaking.

Neutrino Masses and Leptogenesis from R Parity Violation
Ma, E.
Dark Matter in Astro- and Particle Physics, 2001/1/1, 448
Abstract - Not Available

Neutrino Masses and Mixing from Supersymmetric Inflation
Lazarides, G.
COSMO-97, First International Workshop on Particle Physics and the Early Universe, 1998/1/1, 327
Abstract - Not Available

Neutrino Masses and Mixings: Big Bang and Supernova Nucleosynthesis and Neutrino Dark Matter
Fuller, G. M.
AIP Conf. Proc. 478: COSMO-98, 1999/1/1, 448
Abstract - Not Available

Neutrino mixing and oscillations in 1999 and beyond
Petcov, S. T.
Weak Interactions and Neutrinos, 2000/1/1, 305
Abstract - Not Available

Neutrino Mixing from Neutrino Oscillation Data
Bilenky, S. M.; Giunti, C.; Grimus, W.
Particle and Nuclear Physics, 1998/1/1, 219
Abstract - Not Available

Neutrino mixing schemes
Barger, V.; Whisnant, K.
Current aspects of neutrino physics, 2001/1/1, 199
Abstract - Contents: 1. Introduction. 2. Two-neutrino analyses. 3. Global analysis. 4. Consequences for masses and mixings. 5. Long-baseline experiments. 6. Summary and outlook.

Neutrino nucleosynthesis (modern status)
Nadyozhin, D. K.
Cosmic evolution, 2001/1/1, 125
Abstract - Not Available

Neutrino opacities at high density and the protoneutron star evolution
Reddy, S.; Pons, J.; Prakash, M.; Lattimer, J. M.
Stellar Evolution, Stellar Explosions and Galactic Chemical Evolution, 1998/1/1, 585
Abstract - Not Available

Neutrino Oscillation at LSND
Imlay, R. I.
COSMO-97, First International Workshop on Particle Physics and the Early Universe, 1998/1/1, 90
Abstract - Not Available

Neutrino oscillation constraints on neutrinoless double-beta decay
Bilenkey, S. M.; Giunti, C.
Weak Interactions and Neutrinos, 2000/1/1, 190
Abstract - Not Available

Neutrino Oscillation Effects in Indirect Detection of Dark Matter
Fornengo, N.
Dark Matter in Astro- and Particle Physics, 2001/1/1, 659
Abstract - Not Available

Neutrino oscillation experiments at nuclear reactors
Gratta, G.

Weak Interactions and Neutrinos, 2000/1/1, 376
Abstract - Not Available

Neutrino Oscillation Results from Karmen
Zeitnitz, B.; Armbruster, B.; Becker, M.; Benen, A.; Drexlin, G.; Eberhard, V.; Eitel, K.; Gemmeke, H.; Jannakos, T.; Kleifges, M.; Kleinfeller, J.; Oehler, C.; Plischke, P.; Rapp, J.; Reichenbacher, J.; Schnürer, F.; Steidl, M.; Wolf, J.; Bodmann, B. A.; Finckh, E.; Haug, S.; Hößl, J.; Jünger, P.; Kretschmer, W.; Stucken, I.; Eichner, C.; Maschuw, R.; Ruf, C.; Blair, I. M.; Edgington, J. A.; Seligmann, B.; Booth, N. E.
Particle and Nuclear Physics, 1998/1/1, 169
Abstract - Not Available

Neutrino Oscillation Search in CHORUS and NOMAD
Herin, J.
COSMO-97, First International Workshop on Particle Physics and the Early Universe, 1998/1/1, 85
Abstract - Not Available

Neutrino Oscillation Searches at CHORUS
Ricciardi, S.
AIP Conf. Proc. 415: Beyond the Standard Model. From Theory to Experiment, 1998/1/1, 316
Abstract - Not Available

Neutrino Oscillations
Pakvasa, S.
AIP Conf. Proc. 444: Particle Physics and Cosmology, First Tropical Workshop, 1998/1/1, 17
Abstract - Not Available

Neutrino oscillations and blazars
Mannheim, Karl
Astroparticle Physics, 11, 1999/6/1, 49-57
Abstract - Three independent predictions follow from postulating the existence of protons co-accelerated with electrons in extragalactic jets (i) multi-TeV gamma ray emission from nearby blazars, (ii) extragalactic cosmic ray protons up to $\sim 10^{20}$ eV, and (iii) extragalactic neutrinos up to $\sim 5 \times 10^{18}$ eV. Recent gamma ray observations of Mrk 421 and Mrk 501 employing the air-Cerenkov technique are consistent with the predicted gamma ray spectrum, if one corrects for pair attenuation on the infrared background. Prediction (ii) is consistent with cosmic ray data, if one requires that jets are responsible for a at least a sizable fraction of the extragalactic gamma ray background. With cubic kilometer neutrino telescopes, it will be possible to test (iii), although the muon event rates are rather low. Neutrino oscillations can increase the event rate by inducing tau-cascades removing the so-called Earth shadowing effect.

Neutrino Oscillations and Cosmology
Dolgov, A. D.
Current Topics in Astrofundamental Physics: the Cosmic Microwave Background, 2001/1/1, 565
Abstract - Not Available

Neutrino Oscillations and Elementarity in Leptonic Matter
Rajpoot, Subhash
Particles, Strings and Cosmology, 2000/1/1, 429

Neutrino Oscillations and Gamma-Ray Bursts
Kluzniak, W.
Astrophysical Journal, 508, 1998/11/1, L29-L31
Abstract - If the ordinary neutrinos oscillate into a sterile flavor in a manner consistent with the Super-Kamiokande data on the zenith-angle dependence of atmospheric mu-neutrino flux, an energy sufficient to power a typical cosmic gamma-ray burst (GRB) ($\sim 10^{52}$ ergs) can be carried away from the source by sterile neutrinos and deposited in a region relatively free of baryons. Hence, ultrarelativistic bulk motion (required by the theory and observations of GRBs and their afterglows) can be achieved in the vicinity of plausible sources of GRBs. Oscillations between sterile and ordinary neutrinos might, therefore, provide a solution to the "baryon-loading problem" in the theory of GRBs.

Neutrino oscillations and the solar neutrino
  problem
  Haxton, W. C.
  Current aspects of neutrino physics,
  2001/1/1, 65
  Abstract - Contents: 1. Introduction. 2.
  Open questions in neutrino physics. 3. The
  standard solar model. 4. Solar neutrino
  experiments and their implications. 5.
  Neutrino oscillations. 6. The Mikheyev-
  Smirnov-Wolfenstein mechanism. 7.
  Outlook.

Neutrino Oscillations in Magnetized Media and
  Implications for the Pulsar Velocity Puzzle
  Grasso, D.
  AIP Conf. Proc. 415: Beyond the Standard
  Model. From Theory to Experiment,
  1998/1/1, 322
  Abstract - Not Available

Neutrino oscillations in the early universe: how
  can large lepton asymmetry be generated?
  Dolgov, A. D.; Hansen, S. H.; Pastor, S.;
  Semikoz, D. V.
  Astroparticle Physics, 14, 2000/9/1, 79-90
  Abstract - The lepton asymmetry that could
  be generated in the early universe through
  oscillations of active to sterile neutrinos is
  calculated (almost) analytically for small
  mixing angles, $\sin 2\theta < 10^{-2}$. It is shown that for a mass
  squared difference,
  $\delta m^2 = -1$
  eV^2 it may rise at most by
  six orders of magnitude from the initial
  "normal" value /~10^{-10}, as
  the back-reaction from the refraction index
  terminates this rise while the asymmetry is
  still small. Only for very large mass
  differences,
  $|\delta m^2| \sim 10^9$ eV^2, the lepton
  asymmetry could reach a significant
  magnitude exceeding 0.1.

Neutrino Oscillations: a phenomenological
  overview
  Fogli, G. I.; Lisi, F.; Montanino, O.;
  Scioscia, G.
  Dark matter : proceedings of DM97, 1st
  Italian conference on dark matter, Trieste,
  December 9-11, 1997 / editor, Paolo

Salucci. Firenze, Italy : Studio Editoriale
  Fiorentino, c1998, p. 89., 1998/1/1, 89
  Abstract - Not Available

Neutrino oscillations: a source of Goldstone
  fields and consequences for supernovae
  Bento, L.
  Lepton and Baryon Number Violation in
  Particle Physics, neral Astrophysics and
  Cosmology, 1999/1/1, 487
  Abstract - Not Available

Neutrino oscillations: accelerator experiments
  Maschuw, R.
  Weak Interactions and Neutrinos,
  2000/1/1, 426
  Abstract - Not Available

Neutrino Pair Annihilation above a Kerr Black
  Hole with the Accretion Disk
  Asano, Katsuaki
  Gamma-ray Bursts in the Afterglow Era,
  2001/1/1, 318
  Abstract - We investigate the relativistic
  effects on the energy deposition rate via
  neutrino pair annihilation near the rotation
  axis of a Kerr black hole. For the disk with
  a temperature gradient, the energy
  deposition rate for a small inner radius of
  the accretion disk is smaller than that
  estimated by neglecting the relativistic
  effects. The relativistic effects, especially
  for a large Kerr parameter, a, play a
  negative role in avoiding the baryon
  contamination problem in gamma-ray
  bursts.

Neutrino Pair Annihilation in the Gravitation of
  Gamma-Ray Burst Sources
  Asano, Katsuaki; Fukuyama, Takeshi
  Astrophysical Journal, 531, 2000/3/1, 949-
  955
  Abstract - We study semianalytically the
  gravitational effects on neutrino pair
  annihilation near the neutrinosphere and
  around the thin accretion disk. For the disk
  case, we assume that the accretion disk is
  isothermal and that the gravitational field is
  dominated by the Schwarzschild black
  hole. General relativistic effects are studied
  only near the rotation axis. The energy
  deposition rate is enhanced by the effect of
  orbital bending toward the center.

However, the effects of the redshift and gravitational trapping of the deposited energy reduce the effective energy of the gamma-ray burst's source. Although each effect is substantial, the effects partly cancel one another. As a result, the gravitational effects do not substantially change the energy deposition rate for either the spherically symmetric case or the disk case.

Neutrino Physics
Akhmedov, E. Kh.
Particle Physics, Proceedings of the 1999 Summer School, held in Trieste, Italy, 21 June - 9 July, 1999. The ICTP Series in Theoretical Physics. Edited by G. Senjanovic and A. Yu. Smirnov. Singapore: World Scientific Press, 2000., p.103, 2000/1/1, 103
Abstract - Not Available

Neutrino Physics
Peccei, R. D.
Particles and Fields, Eighth Mexican School, 1999/1/1, 80
Abstract - Not Available

Neutrino Physics and Astrophysics
Akhmedov, E. K.
New worlds in astroparticle physics, 2001/1/1, 3
Abstract - Not Available

Neutrino Physics and Astrophysics
Suzuki, Y.; Totsuka, Y.
Neutrino Physics and Astrophysics, Proceedings of the XVIIIth International Conference on Neutrino Physics and Astrophysics, Takayama, Japan, 4-9 June 1998 Edited by Y. Suzuki and Y. Totsuka. Elsevier Science, 1999., 1999/1/1
Abstract - The scientific program of these important proceedings was arranged to cover most of the field of neutrino physics. In light of the rapid growth of interest stimulated by new interesting results from the field, more than half of the papers presented here are related to the neutrino mass and oscillations, including atmospheric and solar neutrino studies. Neutrino mass and oscillations could imply the existence of a mass scale many orders

of magnitudes higher than presented in current physics and will probably guide scientists beyond the standard model of particle physics.

Neutrino Physics and Astrophysics
Vignaud, D.
New Worlds in Astroparticle Physics, 1998/1/1, 3
Abstract - Not Available

Neutrino Physics and the Primordial Elemental Abundances
Cardall, C. Y.
Eighteenth Texas Symposium on Relativistic Astrophysics, 1998/1/1, 245
Abstract - Not Available

Neutrino Physics with MACRO Detector
Bernardini, P.
New Worlds in Astroparticle Physics, 1998/1/1, 200
Abstract - Not Available

Neutrino Physics with Thermal Detectors
Nucciotti, A. et al.
New Worlds in Astroparticle Physics II, 1999/1/1, 148
Abstract - Not Available

Neutrino Physics, Science of the New Millennium
Li, Xue-Qian
Symposium on the Frontiers of Physics at Millenium, 2001/1/1, 148
Abstract - Not Available

Neutrino Physics: Status and Prospect
Valle, J. W. F.
Dark Matter in Astro- and Particle Physics, 2001/1/1, 379
Abstract - Not Available

Neutrino Process Contributions to LiBeB Nucleosynthesis
Hartmann, D.; Myers, J.; Woosley, S.; Hoffman, R.; Haxton, W.
ASP Conf. Ser. 171: LiBeB Cosmic Rays, and Related X- and Gamma-Rays, 1999/1/1, 235
Abstract - Not Available

Neutrino Production by the Moon
Wilson, T. L.
Lunar and Planetary Institute Conference
Abstracts, 29, 1998/3/1, 1114
Abstract - Not Available.

Neutrino propagation through dense matter
Naumov, Vadim A.; Perrone, Lorenzo
Astroparticle Physics, 10, 1999/3/1, 239-
252
Abstract - We propose a simple approach
to solving the transport equation for high-
energy neutrinos in dense and thick media.
Illustrative results obtained from some
specific models for the initial spectra of
muon neutrinos and antineutrinos
propagating through a normal cold medium
are presented.

Neutrino propagation through fluctuating media
Burgess, C. P.
Weak Interactions and Neutrinos,
2000/1/1, 207
Abstract - Not Available

Neutrino Puzzles and Their Implications for the
Nature of New Physics
Mohapatra, R. N.
COSMO-97, First International Workshop
on Particle Physics and the Early Universe,
1998/1/1, 71
Abstract - Not Available

Neutrino Puzzles and Their Implications for the
Nature of New Physics
Mohapatra, R. N.
Particle and Nuclear Physics, 1998/1/1, 55
Abstract - Not Available

Neutrino Reactions in Nuclei and Neutrino
Backgrounds
Mintz, S. L.; Pourkaviani, M.
Confluence , of Cosmology, Massive
Neutrinos, Elementary Particles, and
Gravitation, 1999/1/1, 71
Abstract - Not Available

Neutrino signals from WIMP annihilation
Fornengo, N.
Weak Interactions and Neutrinos,
2000/1/1, 253
Abstract - Not Available

Neutrino Spectroscopy with Karmen
Maschuw, R.
Particle and Nuclear Physics, 1998/1/1,
183
Abstract - Not Available

Neutrino spin flip in system of magnetic
barriers
Apostolovska, G. et al.
IAU Symposium, 203, 2000/1/1, 59
Abstract - If a neutrino has a non-vanishing
magnetic moment it's helicity can be
flipped when it passes through a region
with a magnetic field which has a
component perpendicular to the direction
of motion of the neutrino. We have
explored the tunneling of a relativistic
neutral spin 1/2 particle (Dirac neutrino)
through a finite number of magnetic
barriers (wells). Using transfer matrix
tehnique we have calculated the
transmission coefficients and polarization
phase change as a function of the incident
angle and energy of the particle.

Neutrino telescopes in Antarctica
Adams, J.
Publications of the Astronomical Society
of Australia, 17, 2000/4/1, 13-17
Abstract - It is hoped that in the near future
neutrino astronomy will reach throughout
and beyond our galaxy and make
measurements relevant to cosmology,
astrophysics, cosmic-ray and particle
physics. The construction of a high-energy
neutrino telescope requires a huge volume
of very transparent, deeply buried material
such as ocean water or ice, which acts as
the medium for detecting the particles. I
will describe two experiments using
Antarctic ice as this medium: the
AMANDA experiment employing
photomultiplier tubes and RICE utilizing
radio receivers.

Neutrino Transport and Large-Scale
Convection in Core-Collapse Supernovae
Guidry, M.
Neutrino Physics and Astrophysics,
1998/1/1, 115
Abstract - Not Available

Neutrino Transport in Strongly Magnetized Proto-Neutron Stars and the Origin of Pulsar Kicks: The Effect of Asymmetric Magnetic Field Topology
Lai, Dong; Qian, Yong-Zhong
Astrophysical Journal, 505, 1998/10/1, 844-853
Abstract - In proto-neutron stars with strong magnetic fields, the cross section for nu_e (nu&d1;_e) absorption on neutrons (protons) depends on the local magnetic field strength resulting from the quantization of energy levels for the e^- (e^+) produced in the final state. If the neutron star possesses an asymmetric magnetic field topology in the sense that the magnitude of magnetic field in the north pole is different from that in the south pole, then asymmetric neutrino emission may be generated. We calculate the absorption cross sections of nu_e and nu&d1;_e in strong magnetic fields as a function of the neutrino energy. These cross sections exhibit oscillatory behaviors that occur because new Landau levels for the e^- (e^+) become accessible as the neutrino energy increases. By evaluating the appropriately averaged neutrino opacities, we demonstrate that the change in the local neutrino flux caused by the modified opacities is rather small. To generate appreciable kick velocity ($\sim$300 km s^-1) to the newly formed neutron star, the difference between the field strengths at the two opposite poles of the star must be at least 10^16 G. We also consider the magnetic field effect on the spectral neutrino energy fluxes. The oscillatory features in the absorption opacities give rise to modulations in the emergent spectra of nu_e and nu&d1;_e.

Neutrino Transport in Type II Supernovae Boltzmann Solver vs. Monte Carlo Method
Yamada, Shoichi; Janka, Han-Thomas; Suzuki, Hideyuki
Recent Developments in Theoretical and Experimental General Relativity, Gravitation, and Relativistic Field Theories, 1999/1/1, 1468
Abstract - Not Available

Neutrino Transport in Type II Supernovae: Boltzmann Solver VS Monte Carlo Method
Yamada, S.; Janka, H. T.; Suzuki, H.
IAU Symp. 188: The Hot Universe, 188, 1998/1/1, 247
Abstract - The Collapse-driven supernova is an explosive phenomenon which occurs under the extremely high density, high temperature and strong gravity condition. The best probe for this object is neutrinos. Many researchers in this field nowadays think neutrinos are playing a crucial role in the supernova explosion. In addition, close study of the information neutrinos carry away from the supernova core will provide us with in-depth understanding of the nature of nuclear matter and neutrino response in the high density and high temperature medium. However, since neutrino transport in supernova simulations has been treated thus far mainly by some approximate methods like the multi-group flux limited diffusion scheme (MGFLD), a more sophisticated method is required. In this contributed paper we show the performance of the Boltzmann solver which is recently developed by us, comparing with the Monte Carlo simulations and MGFLD. Some applications will be also addressed.

Neutrino transport in type II supernovae: Boltzmann solver vs. Monte Carlo method
Yamada, Shoichi; Janka, Hans-Thomas; Suzuki, Hideyuki
Astronomy and Astrophysics, 344, 1999/4/1, 533-550
Abstract - We have coded a Boltzmann solver based on a finite difference scheme (S_N method) aiming at calculations of neutrino transport in type II supernovae. Close comparison between the Boltzmann solver and a Monte Carlo transport code has been made for realistic atmospheres of post bounce core models under the assumption of a static background. We have also investigated in detail the dependence of the results on the numbers of radial, angular, and energy grid points and the way to discretize the spatial advection term which is used in the Boltzmann solver. A general relativistic

calculation has been done for one of the models. We find good overall agreement between the two methods. This gives credibility to both methods which are based on completely different formulations. In particular, the number and energy fluxes and the mean energies of the neutrinos show remarkably good agreement, because these quantities are determined in a region where the angular distribution of the neutrinos is nearly isotropic and they are essentially frozen in later on. On the other hand, because of a relatively small number of angular grid points (which is inevitable due to limitations of the computation time) the Boltzmann solver tends to slightly underestimate the flux factor and the Eddington factor outside the (mean) "neutrinosphere" where the angular distribution of the neutrinos becomes highly anisotropic. As a result, the neutrino number (and energy) density is somewhat overestimated in this region. This fact suggests that the Boltzmann solver should be applied to calculations of the neutrino heating in the hot-bubble region with some caution because there might be a tendency to overestimate the energy deposition rate in disadvantageous situations. A comparison shows that this trend is opposite to the results obtained with a multi-group flux-limited diffusion approximation of neutrino transport.

Neutrino-driven Jets and Rapid-Process Nucleosynthesis
Nagataki, Shigehiro
Astrophysical Journal, 551, 2001/4/1, 429-438
Abstract - We have studied whether the jet in a collapse-driven supernova can be a key process for the rapid-process (r-process) nucleosynthesis. We have examined the features of a steady, subsonic, and rigidly rotating jet in which the centrifugal force is balanced by the magnetic force. As for the models in which the magnetic field is weak and angular velocity is small, we found that the r-process does not occur because the final temperature is kept too high and the dynamical timescale becomes too long

when the neutrino luminosities are set to be high. Even if the luminosities of the neutrinos are set to be low, which results in the low final temperature, we found that the models do not give a required condition to produce the r-process matter. Furthermore, the amount of the mass outflow seems to be too little to explain the solar system abundance ratio in such low-luminosity models. As for the models in which the magnetic field is strong and angular velocity is large, we found that the entropy per baryon becomes too small and the dynamical timescale becomes too long. This tendency is, of course, a bad one for the production of the r-process nuclei. As a conclusion, we have to say that it is difficult to cause a successful r-process nucleosynthesis in the jet models in this study.

Neutrino-driven supernovae: Boltzmann neutrino transport and the explosion mechanism
Messer, O. E. B.; Mezzacappa, A.; Bruenn, S. W.; Guidry, M. W.
Stellar Evolution, Stellar Explosions and Galactic Chemical Evolution, 1998/1/1, 563
Abstract - Not Available

Neutrino-Electron Scattering as a Probe of the Electroweak Gauge Structure
Miranda, O. G.; Semikoz, V.; Valle, J. W. F.
AIP Conf. Proc. 415: Beyond the Standard Model. From Theory to Experiment, 1998/1/1, 340
Abstract - Not Available

Neutrino-induced Fission and r-Process Nucleosynthesis
Qian, Y.-Z.
Astrophysical Journal, 569, 2002/4/1, L103-L106
Abstract - An r-process scenario with fission but without fission cycling is considered to account for the observed abundance patterns of neutron-capture elements in ultra-metal-poor stars. It is proposed that neutrino reactions play a crucial role in inducing the fission of the progenitor nuclei after the r-process freezes

out in Type II supernovae. To facilitate neutrino-induced fission, the proposed r-process scenario is restricted to occur in a low-density environment, such as the neutrino-driven wind from the neutron star. Further studies to develop this scenario are emphasized.

Neutrino-Induced Synthesis of ^{7}LI in He-Shell
Nadyozhin, D. K.
LNP Vol. 287: Proceedings of the 9th workshop on Nuclear Astrophysics, 1998/1/1, 119
Abstract - Not Available

Neutrinoless Double Beta Decay Potential in a Large Mixing Angle World
Klapdor-Kleingrothaus, H. V. et al.
Dark Matter in Astro- and Particle Physics, 2001/1/1, 420
Abstract - Not Available

Neutrinoless double beta decay with Xe-136 in BOREXINO and the BOREXINO Counting Test Facility
Caccianiga, B.; Giammarchi, M. G.
Astroparticle Physics, 14, 2000/8/1, 15-31
Abstract - This article discusses the methods and sensitivity for a double beta decay experiment based on the Xe-136 candidate for BOREXINO or the BOREXINO Counting Test Facility. Different background assumptions and experimental configurations are studied, assuming a data obtaining period of one year. The related experimental problems are discussed, and summary tables containing the sensitivity estimates for the various configurations are presented.

Neutrino-pair bremsstrahlung by electrons in neutron star crusts
Kaminker, A. D.; Pethick, C. J.; Potekhin, A. Y.; Thorsson, V.; Yakovlev, D. G.
Astronomy and Astrophysics, 343, 1999/3/1, 1009-1024
Abstract - Neutrino-pair bremsstrahlung by relativistic degenerate electrons in a neutron-star crust at densities 10(9) g cm (-3) <~ rho <~ 1.5 x 10(14) g cm(-3) is analyzed. The processes taken into account are neutrino emission due to Coulomb scattering of electrons by atomic nuclei in a Coulomb liquid, and electron-phonon scattering (the phonon contribution) and Bragg diffraction (the static-lattice contribution) in a Coulomb crystal. The static-lattice contribution is calculated including the electron band-structure effects for cubic Coulomb crystals of different types and also for the liquid crystal phases composed of rod- and plate-like nuclei near the bottom of the neutron-star crust (10(14) g cm(-3) <~ rho <~ 1.5 x 10(14) g cm(-3)). The phonon contribution is evaluated with proper treatment of the multi-phonon processes which removes a jump in the neutrino bremsstrahlung emissivity at the melting point obtained in previous works. Generally, bremsstrahlung in the solid phase does not differ significantly from that in the liquid. A comparison of the various neutrino generation mechanisms in neutron star crusts shows that electron bremsstrahlung is among the most important ones.

Neutrino-pair emission in a strong magnetic field
van Dalen, E. N. E.; Dieperink, A. E. L.; Sedrakian, A.; Timmermans, R. G. E.
Astronomy and Astrophysics, 360, 2000/8/1, 549-558
Abstract - We study the neutrino emissivity of strongly magnetized neutron stars due to the charged and neutral current couplings of neutrinos to baryons in strong magnetic fields. The leading order neutral current process is the one-body neutrino-pair bremsstrahlung, which does not have an analogue in the zero field limit. The leading order charged current reaction is the known generalization of the direct Urca processes to strong magnetic fields. While for superstrong magnetic fields in excess of 10^{18} G the direct Urca process dominates the one-body bremsstrahlung, we find that for fields on the order 10^{16} - 10^{17} G and temperatures a few times 10^{9} K the one-body bremsstrahlung is the dominant process. Numerical computation of the resulting emissivity, based on a simple parametrization of the equation of state of the npe-matter in a strong magnetic field,

shows that the emissivity of this reaction is of the same order of magnitude as that of the modified Urca process in the zero field limit.

Neutrinos
Pakvasa, S.
High Energy Physics and Cosmology, 1997 Summer School, 1998/1/1, 371
Abstract - Not Available

Neutrinos after Takayama
Vannucci, F.
Abstracts of the 19th Texas Symposium on Relativistic Astrophysics and Cosmology, held in Paris, France, Dec. 14-18, 1998. Eds.: J. Paul, T. Montmerle, and E. Aubourg (CEA Saclay)., 1998/12/1, 374
Abstract - The SuperKamiokande collaboration has given evidence of oscillations for neutrinos of atmospheric origin. The result will be discussed and the consequences for neutrino masses will be sketched.

Neutrinos and Core Collapse Supernovae
Janka, H.-T.
Particle and Nuclear Physics, 1998/1/1, 31
Abstract - Not Available

Neutrinos and Extra Dimensions
Caldwell, D. O.
Identification of Dark Matter, 2001/1/1, 513
Abstract - Not Available

Neutrinos and Muons
Kielczewska, D.
AIP Conf. Proc. 516: 26th International Cosmic Ray Conference, ICRC XXVI , 26, 2000/1/1, 225
Abstract - Not Available

Neutrinos and Nuclear Responses in Nuclear DOUBLE-ß and INVERSE-ß Processes
Ejiri, H.
Particle and Nuclear Physics, 1998/1/1, 307
Abstract - Not Available

Neutrinos and Supermassive Stars: Prospects for Neutrino Emission and Detection
Shi, Xiangdong; Fuller, George M.
Astrophysical Journal, 503, 1998/8/1, 307
Abstract - We calculate the luminosity and energy spectrum of the neutrino emission from electron-positron pair annihilation during the collapse of a supermassive star (M >~ 5 x 104 M&sun;). We then estimate the cumulative flux and energy spectrum of the resulting neutrino background as a function of the abundance and redshift of supermassive stars and the efficiency of these objects in converting gravitational energy into neutrino energy. We estimate the expected signal in some of the new generation of astrophysical neutrino detectors from both a cumulative background of supermassive stars and single collapse events associated with these objects.

Neutrinos and supernova theory.
Burrows, A.; Young, T.
Physics Reports, 333, 2000/1/1, 63-75
Abstract - Electronic Article Available from Elsevier Science.

Neutrinos from active galactic nuclei
Schuster, C.; Pohl, M.; Schlickeiser, R.
Astronomische Gesellschaft Meeting Abstracts, 15, 1999/1/1
Abstract - The neutrino production in jets of active galactic nuclei resulting from inelastic proton-proton collisions is calculated. The jets are modelled as a highly relativistic plasma that interacts with ambient matter by sweeping up interstellar gas. It has been shown that the swept-up matter may be quickly isotropised in the blast wave frame by a relativistic two-stream instability, which provides relativistic protons and electrons in the jet. At the same time the blast wave decelerates because of momentum conservation. Inelastic interactions of the relativistic protons in the blast wave plasma generate neutrinos via pion decay and muon decay. The resulting neutrino fluxes are compared with the atmospheric neutrino fluxes.

Neutrinos from active galactic nuclei as a diagnostic tool
Schuster, C.; Pohl, M.; Schlickeiser, R.
Astronomy and Astrophysics, 382,

2002/2/1, 829-837
Abstract - Active galactic nuclei (AGN) are known as sources of high energy gamma -rays. The emission probably results from non-thermal radiation of relativistic jets belonging to the AGN. Earlier investigations of these processes have suggested that neutrinos are among the radiation products of the jets and may be used to discriminate between hadrons and leptons as primary particles for the production of the high energy emission. Our calculation of the high energy neutrino emission from the jets of AGN is based on a recently published model for gamma -ray production by a collimated, relativistic blast wave, in which the spectral evolution of energetic particles is determined by the interplay between the particle injection by sweep-up of the interstellar medium, the energy losses through radiation, and diffusive escape. It is important to note that the swept-up interstellar particles retain their relative velocities with respect to the jet plasma, but get isotropised in the jet rest frame by self-excited Alfvénic turbulence. The bulk of the neutrino emission is expected in the energy range between 100 GeV and a few TeV. It is shown that the neutrino flux is correlated with the flux of TeV gamma -rays. This allows to distinctly search for neutrino emissions from the jets of AGN by using the TeV gamma -ray light curves to drastically reduce the temporal and spatial parameter space. Given the observed TeV photon fluxes from nearby BL Lacs the neutrino signal from AGN may be detectable with future neutrino observatories as least as sensitive as IceCube.

Neutrinos from AGN
Kazanas, D.
High Energy Gamma-Ray Astronomy, 2001/1/1, 370

Neutrinos from Blazars and the Differences between FRI and FRII Radio Galaxies
Dermer, C. D.; Atoyan, A.; Boettcher, M.
American Astronomical Society Meeting, 199, 2001/12/1
Abstract - Nonthermal proton acceleration in galaxy jets produce neutrinos through photomeson production. Sources with intense surrounding fields should be strong neutrino sources. Neutrons and high-energy photons formed through photomeson processes will transport inner jet energy far from the black hole engine to power distant jets, hot spots and lobes, as observed at X-ray energies with Chandra. This effect may account for differences between the FRI and FRII classes of radio galaxies that display distinct morphologies.

Neutrinos from Cosmic Ray Interactions and Relativistic Astrophysical Sources
Protheroe, Ray
ASP Conf. Ser. 241: The 7th Taipei Astrophysics Workshop on Cosmic Rays in the Universe, 2001/1/1, 93
Abstract - Not Available

Neutrinos from Early-Phase, Pulsar-driven Supernovae
Beall, J. H.; Bednarek, W.
Astrophysical Journal, 569, 2002/4/1, 343-348
Abstract - Neutron stars, just after their formation, are surrounded by expanding, dense, and very hot envelopes that radiate thermal photons. Iron nuclei can be accelerated in the wind zones of such energetic pulsars to very high energies. These nuclei photodisintegrate, and their products lose energy efficiently in collisions with thermal photons and with the matter of the envelope, mainly via pion production. When the temperature of the radiation inside the envelope of the supernova drops below $\sim 3 \times 10^6$ K, these pions decay before losing energy and produce high-energy neutrinos. We estimate the flux of muon neutrinos emitted during such an early phase of the pulsar-supernova envelope interaction. We find that a 1 km_2 neutrino detector should be able to detect neutrinos above 1 TeV within about 1 yr after the explosion from a supernova in our Galaxy. This result holds if these pulsars are able to efficiently accelerate nuclei to energies $\sim 10^{20}$ eV, as postulated recently by some for models of Galactic acceleration of the extremely high energy cosmic rays (EHE CRs).

Neutrinos from Gamma-ray Bursts
Mannheim, K.
High Energy Gamma-Ray Astronomy,
2001/1/1, 417
Abstract - Not Available

Neutrinos from Protoneutron Stars: Probing
Hot and Dense Matter
Reddy, Sanjay K. Ph.D.
Thesis, 1998/1/1, 3
Abstract - In this work, we elucidate the
role of microphysics in the macrophysical
evolution of a newly-born neutron star,
termed a protoneutron star (PNS). The
roles of strong and weak interactions,
particularly their interdependence through
the equation of state (EOS) and neutrino
opacities, during the early neutrino
emission phase is investigated. We
calculate charged and neutral current weak
interaction rates that determine the
neutrino opacities inside a PNS. We
establish an efficient formalism for
calculating the mean free paths for
interacting, nucleonic matter at arbitrary
degeneracy. First, strong interaction
corrections are incorporated through in-
medium single particle energies, using both
potential and field-theoretical models.
Second, strong and electromagnetic
correlations are calculated using the
Random Phase Approximation and are
found to substantially modify the neutrino
mean free path. In matter containing
hyperons, neutral current tree-level
neutrino-hyperon couplings were
computed for the first time. Together with
known charged current couplings, these
couplings were employed to calculate
neutrino opacities in hyperonic matter.
Simulations of the evolution of a PNS were
performed using different EOS models,
including those with hyperons, to predict
the corresponding neutrino luminosities.
Hyperons considerably soften the EOS and
increase the specific heat of matter during
the late stages of deleptonization. Relative
to nucleons-only case, the central density
and the neutrino opacities during the
cooling phase are significantly larger. This,
however, results in only a marginal
increase in the neutrino diffusion time due
to feedbacks from the larger specific heat
of hyperonic matter. Significant
differences between EOS models with and
without hyperons are manifest in the
emergent neutrino luminosities only for
those cases in which the PNS mass is close
to the maximum mass.

Neutrinos from the Sun
Svoboda, R.; The Super-KAMIOKANDE
Collaboration
AIP Conf. Proc. 516: 26th International
Cosmic Ray Conference, ICRC XXVI , 26,
2000/1/1, 317
Abstract - Not Available

Neutrinos in Cosmology
Gelmini, G.
Particle and Nuclear Physics, 1998/1/1, 1
Abstract - Not Available

Neutrinos in Physics and Astrophysics
Roulet, Esteban
From the Sun to the Great Attractor, 1999
Guanajuato Lectures on Astrophysics
Guanajuato, Mexico 4-11 August 1999,
Edited by D. Pageand J.G. Hirsch, Lecture
Notes in Physics, vol. 556, p.233,
2000/1/1, 233
Abstract - An elementary general overview
of the neutrino physics and astrophysics is
given. We start by a historical account of
the development of our understanding of
neutrinos and how they helped to unravel
the structure of the Standard Model. We
discuss why it is so important to establish
if neutrinos are massive and we introduce
the main scenarios to provide them a mass.
The present bounds and the positive
indications in favor of non-zero neutrino
masses are discussed as well as the major
role they play in astrophysics and
cosmology.

Neutrinos in Physics and Astrophysics
Roulet, Esteban
From the Sun to the Great Attractor,
2000/1/1, 233
Abstract - Not Available

Neutrinos in physics, astrophysics, and
cosmology
Dolgov, A. G.
Memorie della Societa Astronomica

Italiana, 72, 2001/1/1, 823
Abstract - A brief review of neutrino
anomalies in particle physics and of the
role played by neutrinos in cosmology and
astrophysics is presented. The main part of
the talk is dedicated to the impact of
neutrinos and in particular of neutrino
oscillations on BBN and to a possible
spatial variation of primordial abundances.

Neutrinos in the Formation of Heavy Elements
Surman, Rebecca
News Letter of the Astronomical Society
of New York, 5, 1999/8/1, 5
Abstract - Not Available

Neutrinos, cosmology and astrophysics
Gelmini, G.
Weak Interactions and Neutrinos,
2000/1/1, 136
Abstract - Not Available

Neutrinos, Gamma Rays and EUV
ASP Conf. Ser. 241: The 7th Taipei
Astrophysics Workshop on Cosmic Rays
in the Universe, 2001/1/1, 91
Abstract - Not Available

New Directions for New Dimensions: From
Strings to Neutrinos to Axions to
Dienes, Keith R.
Particles, Strings and Cosmology,
2000/1/1, 46
Abstract - Not Available

New era in neutrino physics
Minakata, H.; Yasuda, O.
New Era in Neutrino Physics , 1998/1/1
Abstract - Not Available

New Evidence for Neutrino Degeneracy in the
Early Universe
Mathews, G.; Kajino, T.; Orito, M.
American Astronomical Society Meeting,
197, 2000/12/1
Abstract - We reanalyze cosmological
constraints on the existence of a net
universal lepton asymmetry and neutrino
degeneracy in light of reanalyzed
primordial nucleosynthesis and the recently
reported CMB power spectra from
BOOMERANG and MAXIMA-1. We
explore physically plausible lepton-

asymmetric models with large ν $_{\mu}$
and ν $_{\tau}$ degeneracies together with a
moderate ν $_{e}$ degeneracy.

New Limits on a Diffuse Flux of >= 100 Ee
V Cosmic Neutrinos
Gorham, P. W.; Liewer, K. M.; Naudet, C.
J.; Saltzberg, D. P.; Williams, D.
AIP Conf. Proc. 579: Radio Detection of
High Energy Particles, 2001/1/1, 177
Abstract - Not Available

New Physics at Large Neutrino Telescopes
Moscoso, L.
AIP Conf. Proc. 415: Beyond the Standard
Model. From Theory to Experiment,
1998/1/1, 476
Abstract - Not Available

New Results for Relevant Neutrino Emission in
Superfluid Neutron Star Matter
Pizzochero, Pierre M.
Astrophysical Journal, 502, 1998/8/1, L153
Abstract - We evaluate exactly the
suppression in superfluid neutron star
matter of the neutrino emission due to the
modified Urca process. Corresponding to
realistic parameters for the neutron
superfluid in the core, the reduced
emissivity is found to be 1-3 orders of
magnitude larger than that given by the
approximate expressions used so far in the
literature. We show that this effect has
significant observable consequences on the
evolution of the surface temperature of the
star, whether or not any rapid cooling
occurs in the deepest regions of the core.
These results should be taken into account
in realistic cooling calculations.

New results from the Mainz neutrino mass
experiment
Bonn, J. et al.
Weak Interactions and Neutrinos,
2000/1/1, 160
Abstract - Not Available

New Soudan 2 Results on Neutrino Oscillations
and Cosmic Rays
Marshak, M. L.; The SOUDAN 2
Collaboration
Abstracts of the 19th Texas Symposium on
Relativistic Astrophysics and Cosmology,

held in Paris, France, Dec. 14-18, 1998. Eds.: J. Paul, T. Montmerle, and E. Aubourg (CEA Saclay)., 1998/12/1, 286
Abstract - The Soudan 2 Detector, a 1,000-ton deep underground iron calorimeter designed to search for proton decay has now recorded and analyzed approximately 4 kT-yr of contained event data and 2 times 10^{14} cm^2 -s of deep underground muon data. We report here on new analyses of the atmospheric neutrino contained event data, which estimate Delta m^2 and sin^2 2 theta for neutrino oscillations. We also report on measurements of primary cosmic ray composition from both deep underground multimuon measurements and from combined surface air shower and deep underground muon measurements of ultra high energy cosmic ray events. We also extend previously published data on muons from the direction of Cygnus X-3.

Nonequilibrium Cosmic Neutrions and Nucleosynthesis
Dolgov, A. D.
Frontiers Science Series 23: Black Holes and High Energy Astrophysics, 1998/1/1, 117
Abstract - Not Available

Non-equilibrium Massive Tau-Neutrinos in the Early Universe
Dolgov, A. D.; Hansen, S. H.; Semikoz, D. V.
COSMO-97, First International Workshop on Particle Physics and the Early Universe, 1998/1/1, 136
Abstract - Not Available

Nonequilibrium Neutrinos in Cosmology
Dolgov, A. D.
New Worlds in Astroparticle Physics, 1998/1/1, 184
Abstract - Not Available

Nonlinear Neutrino Interactions with the Stellar Envelope of a Supernova
Dawson, J. M.
New Worlds in Astroparticle Physics II, 1999/1/1, 215
Abstract - Not Available

Nuclear Physics Issues in Neutrino Oscillation Experiments
Singh, S. K.
AIP Conf. Proc. 415: Beyond the Standard Model. From Theory to Experiment, 1998/1/1, 311
Abstract - Not Available

Nuclear Physics Limits to the ^{8}B Solar Neutrino Flux
Mourão, A. M.
New Worlds in Astroparticle Physics, 1998/1/1, 159
Abstract - Not Available

Observation of atmospheric muon neutrinos with AMANDA
Deyoung, Tyce Robert Ph.D.
Thesis, 2001/10/1, 11
Abstract - The Antarctic Muon and Neutrino Detector Array (AMANDA) is designed to detect high energy neutrinos using the three kilometer thick ice cap covering the South Pole as a target and Cherenkov medium. Neutrinos that undergo charged current interactions with nucleons in the ice will produce ultrarelativistic charged leptons, which are detected through their Cherenkov and stochastic radiation by a three dimensional array of phototubes embedded in the ice cap at depths of 1500 to 2000 meters. The background to the observation of neutrinos is the flux of penetrating muons produced in cosmic ray showers in the atmosphere. This flux is approximately one million times the neutrino flux. To reject this background, we look downward, using the Earth to filter out all particles except neutrinos. To demonstrate the correct operation of the detector, we observe atmospheric neutrinos, which are produced in cosmic ray showers in the Northern Hemisphere. The flux, energy spectrum, and angular distribution of these neutrinos are relatively well known, making them a convenient calibration source. This work describes algorithms that have been developed to reconstruct and identify upgoing neutrinos in data recorded during the austral winter of 1997. A total of 204 neutrino candidates are identified, containing less than 10% background from

misreconstructed downgoing muons. The neutrinos observed are found to agree with theoretical predictions of the atmospheric flux within the estimated systematic uncertainties. Limits are placed on high energy neutrino emission from known astronomical sources of very high energy gamma rays.

Observation of atmospheric neutrino events with the AMANDA experiment
Karle, A.; The AMANDA Collaboration
Weak Interactions and Neutrinos, 2000/1/1, 258
Abstract - Not Available

Observation of high-energy neutrinos using Cerenkov detectors embedded deep in Antarctic ice
Andrés, E.; Askebjer, P.; Bai, X.; Barouch, G.; Barwick, S. W.; Bay, R. C.; Becker, K.-H.; Bergström, L.; Bertrand, D.; Bierenbaum, D.; Biron, A.; Booth, J.; Botner, O.; Bouchta, A.; Boyce, M. M.; Carius, S.; Chen, A.; Chirkin, D.; Conrad, J.; Cooley, J.; Costa, C. G. S.; Cowen, D. F.; Dailing, J.; Dalberg, E.; DeYoung, T.; Desiati, P.; Dewulf, J.-P.; Doksus, P.; Edsjö, J.; Ekström, P.; Erlandsson, B.; Feser, T.; Gaug, M.; Goldschmidt, A.; Goobar, A.; Gray, L.; Haase, H.; Hallgren, A.; Halzen, F.; Hanson, K.; Hardtke, R.; He, Y. D.; Hellwig, M.; Heukenkamp, H.; Hill, G. C.; Hulth, P. O.; Hundertmark, S.; Jacobsen, J.; Kandhadai, V.; Karle, A.; Kim, J.; Koci, B.; Köpke, L.; Kowalski, M.; Leich, H.; Leuthold, M.; Lindahl, P.; Liubarsky, I.; Loaiza, P.; Lowder, D. M.; Ludvig, J.; Madsen, J.; Marciniewski, P.; Matis, H. S.; Mihalyi, A.; Mikolajski, T.; Miller, T. C.; Minaeva, Y.; Miocinovic, P.; Mock, P. C.; Morse, R.; Neunhöffer, T.; Newcomer, F. M.; Niessen, P.; Nygren, D. R.; Ögelman, H.; Pérez de los Heros, C.; Porrata, R.; Price, P. B.; Rawlins, K.; Reed, C.; Rhode, W.; Richards, A.; Richter, S.; Martino, J. Rodríguez; Romenesko, P.; Ross, D.; Rubinstein, H.; Sander, H.-G.; Scheider, T.; Schmidt, T.; Schneider, D.; Schneider, E.; Schwarz, R.; Silvestri, A.; Solarz, M.; Spiczak, G. M.; Spiering, C.; Starinsky, N.; Steele, D.; Steffen, P.; Stokstad, R. G.; Streicher, O.; Sun, Q.; Taboada, I.; Thollander, L.; Thon, T.; Tilav, S.; Usechak, N.; Vander Donckt, M.; Walck, C.; Weinheimer, C.; Wiebusch, C. H.; Wischnewski, R.; Wissing, H.; Woschnagg, K.; Wu, W.; Yodh, G.; Young, S.
Nature, 410, 2001/3/1, 441-443
Abstract - Neutrinos are elementary particles that carry no electric charge and have little mass. As they interact only weakly with other particles, they can penetrate enormous amounts of matter, and therefore have the potential to directly convey astrophysical information from the edge of the Universe and from deep inside the most cataclysmic high-energy regions. The neutrino's great penetrating power, however, also makes this particle difficult to detect. Underground detectors have observed low-energy neutrinos from the Sun and a nearby supernova, as well as neutrinos generated in the Earth's atmosphere. But the very low fluxes of high-energy neutrinos from cosmic sources can be observed only by much larger, expandable detectors in, for example, deep water or ice. Here we report the detection of upwardly propagating atmospheric neutrinos by the ice-based Antarctic muon and neutrino detector array (AMANDA). These results establish a technology with which to build a kilometre-scale neutrino observatory necessary for astrophysical observations.

Observation of UHE Neutrino Interactions from Outer Space
Domokos, G.; Kovesi-Domokos, S.
AIP Conf. Proc. 433: Workshop on Observing Giant Cosmic Ray Air Showers From >10(20) eV Particles From Space, 1998/1/1, 390
Abstract - Not Available

Observing the birth of supermassive black holes with the planned ICECUBE neutrino detector.
Shi, X.; Fuller, G. M.
Physical Review Letters, 81, 1998/1/1, 5722-5725
Abstract - Not Available

On Steady State Neutrino-heated Ultrarelativistic Winds from Compact Objects
Pruet, Jason; Fuller, George M.; Cardall, Christian Y.
Astrophysical Journal, 561, 2001/11/1, 957-963
Abstract - We study steady state winds from compact objects in the regime where the wind velocity at infinity is ultrarelativistic. This may have relevance to some models of gamma-ray bursts (GRBs). Particular attention is paid to the case in which neutrinos provide the heating. Unless the neutrino luminosity is very large, $L > 10^{54}$ ergs s^{-1}, the only allowed steady state solutions are those where energy deposition is dominated by neutrino-antineutrino annihilation at the sonic point. In this case, the matter temperature near the neutron star surface is low, less than 1 MeV for typical neutrino luminosities. This is in contrast to the case for subrelativistic winds discussed in the context of supernovae, where the matter temperature near the neutron star approximates the temperature characterizing the neutrinos. We also investigate the setting of the neutron-to-proton ratio (N/P) in these winds and find that only for large (>10 MeV) electron-neutrino or electron-antineutrino temperatures is N/P entirely determined by neutrino capture. Otherwise, N/P retains an imprint of conditions in the neutron star.

On the detection of neutrino oscillations with Planck surveyor
Popa, L.; Burigana, C.; Finelli, F.; Mandolesi, N.
Astronomy and Astrophysics, 363, 2000/11/1, 825-836
Abstract - The imprint of neutrino oscillations on the Cosmic Microwave Background (CMB) anisotropy and polarization power spectra is evaluated in a Lambda CHDM model with two active neutrino flavors, consistent with the structure formation models and the atmospheric neutrino oscillations data. By using the Fisher information matrix method we find that the neutrino oscillations could be detected by Planck surveyor if the present value of the lepton asymmetry $L_{\ν} > = 4.5$ x 10^{-3}, the difference of the neutrino squared masses Delta $m^2 > = 9.7$ x 10^{-3} eV^2 and the vacuum mixing angle sin^2 2θ$_0$ $> = 0.13$, showing the existence of a significant overlap between the region of the oscillation parameter space that can be measured by Planck surveyor and that implied by the atmospheric neutrino oscillations data.

On the detection of ultra high energy neutrinos with the Auger observatory
Capelle, K. S.; Cronin, J. W.; Parente, G.; Zas, E.
Astroparticle Physics, 8, 1998/4/1, 321-328
Abstract - We show that the Auger Air Shower Array has the potential to detect neutrinos of energies in the 10^19 eV range through horizontal air showers. Assuming some simple conservative trigger requirements, we obtain the acceptance for horizontal air showers as induced by high energy neutrinos by two alternative methods and we then give the expected event rates for a variety of neutrino fluxes as predicted in different models which are used for reference.

On the Energy of Neutrinos from Gamma-Ray Bursts
Vietri, Mario
Astrophysical Journal, 507, 1998/11/1, 40-45
Abstract - Ultrahigh-energy protons that are accelerated at the shocks, causing gamma-ray bursts, photoproduce pions and then neutrinos in situ. I consider here the sources of losses in this process, namely, adiabatic and synchrotron losses by both pions and muons. When the shocks under consideration are external, i.e., when they are between the ejecta and the surrounding interstellar medium, I show that neutrinos produced by pion decay are unaffected by losses; those produced by muon decay, in the strongly beamed emission required by afterglow observations of GRB 971214, are limited in energy, but still exceed 10^19 eV.

On the Formation of Dark Matter Balls
Composed of Degenerate, Self-gravitating
Neutrinos
Tsiklauri, David; Viollier, Raoul D.
Astrophysical Journal, 501, 1998/7/1, 486
Abstract - We propose a novel scenario for
the formation and time evolution of dark
matter balls composed of degenerate, self-
gravitating tau -neutrinos and
antineutrinos. This model is based on the
interplay of two competing processes:
annihilation of the particle-antiparticle
pairs via weak interaction and spherical
(Bondi) accretion of these particles. Our
model can account for the formation of
dark matter balls, which could play a role
similar to that of supermassive (~109
M&sun;) black holes in active galactic
nuclei if the neutrino mass is in the range
10-25 keV/c2.

On the Mass of the Neutrino
Teller, Edward
Confluence of Cosmology, Massive
Neutrinos, Elementary Particles, and
Gravitation, 1999/1/1, 3
Abstract - Not Available

On the Neutrino Flux from Gamma-Ray Bursts
Guetta, D.; Spada, M.; Waxman, E.
Astrophysical Journal, 559, 2001/9/1, 101-
109
Abstract - Observations imply that
γ-ray bursts (GRBs) are produced
by the dissipation of the kinetic energy of a
highly relativistic fireball. Photomeson
interactions of protons with γ-rays
within the fireball dissipation region are
expected to convert a significant fraction of
fireball energy to greater than 10^{14} eV
neutrinos. We present an analysis of the
internal shock model of GRBs, where
production of synchrotron photons and
photomeson neutrinos are self-consistently
calculated, and show that the fraction of
fireball energy converted to high-energy
neutrinos is not sensitive to uncertainties in
fireball model parameters, such as the
expansion Lorentz factor and characteristic
variability time. This is due, in part, to the
constraints imposed on fireball parameters
by observed GRB characteristics and, in
part, to the fact that, for parameter values

for which the photomeson optical depth is
high (implying high proton energy loss to
pion production), neutrino production is
suppressed by pion and muon synchrotron
losses. The neutrino flux is, therefore,
expected to be correlated mainly with the
observed γ-ray flux.

On the Neutrino Flux from Gamma-Ray Bursts
Guetta, D.; Spada, M.; Waxman, E.
Gamma-ray Bursts in the Afterglow Era,
2001/1/1, 272
Abstract - Not Available

On the possibility of radar echo detection of
ultra-high energy cosmic ray- and
neutrino-induced extensive air showers
Gorham, P. W.
Astroparticle Physics, 15, 2001/4/1, 177-
202
Abstract - We revisit and extend the
analysis supporting a 60-year-old
suggestion that cosmic rays air showers
resulting from primary particles with
energies above 10^{18} eV should be
straightforward to detect with radar
ranging techniques, where the radar echoes
are produced by scattering from the
column of ionized air produced by the
shower. The idea has remained curiously
untested since it was proposed, but if our
analysis is correct, such techniques could
provide a significant alternative approach
to air shower detection in a standalone
array with high duty cycle, and might
provide highly complementary
measurements of air showers detected in
existing and planned ground arrays such as
the Fly's Eye or the Auger Project. The
method should be particularly sensitive to
showers that are transverse to and
relatively distant from the detector, and is
thus effective in characterizing penetrating
horizontal showers such as those that might
be induced by ultra-high energy neutrino
primaries.

On the relation of extragalactic cosmic ray and
neutrino fluxes
Rachen, J. P.; Mannheim, K.; Protheroe, R.
Abstracts of the 19th Texas Symposium on
Relativistic Astrophysics and Cosmology,
held in Paris, France, Dec. 14-18, 1998.

Eds.: J. Paul, T. Montmerle, and E. Aubourg (CEA Saclay)., 1998/12/1, 641
Abstract - We discuss how and to which extent UHE cosmic ray and neutrino fluxes from extragalactic sources are related to each other and can be used to apply mutual limits. We put particular emphasis on the ejection of cosmic rays from Fermi-accelerating sources and the effect of cosmological evolution. We show that existing models of blazar neutrino production are not necessarily in conflict with UHE cosmic ray data.

On the resonant spin flavor precession of the neutrino in the sun
Derkaoui, J.; Tayalati, Y.
Astroparticle Physics, 14, 2001/1/1, 351-363
Abstract - This work deals with the possible solution of the solar neutrino problem in the framework of the resonant neutrino spin-flavor precession scenario. The event rate results from the solar neutrino experiments as well as the recoil electron energy spectrum from SuperKamiokande are used to constrain the free parameters of the neutrino in this model (Δm^2 and μ_{ν}). We consider two kinds of magnetic profiles inside the sun. For both cases, a static and a twisting field are discussed.

On the Solar-Cycle Modulation of the Homestake Solar Neutrino Capture Rate and the Shuffle Test
Walther, Guenther
Astrophysical Journal, 513, 1999/3/1, 990-996
Abstract - There exists no significant correlation between the Homestake neutrino data up to run 133 and the monthly sunspot number, according to a test that is based on certain optimality properties for this type of problem. It is argued that priorly reported highly significant results for segments of the data are due to a statistical fallacy: the usual methods for evaluating the significance of common tests for correlation are not applicable in the sunspot-neutrino context. Moreover, an appropriate evaluation of these tests gives results that are compatible with the hypothesis of no correlation. Some new methods are introduced for assessing the significance of common measures of correlation in a time series setting, with a special emphasis on the Spearman rank correlation coefficient.

Optical Confirmation of a Neutrino-Detected Supernova
Robinson, L. J.; Roth, J.; Sky Publishing Corp. Team
American Astronomical Society Meeting, 195, 1999/12/1
Abstract - The next nearby supernova is likely to be first detected by neutrino observatories. At best, they will determine its position to within some tens of square degrees. Since the supernova has a 50:50 chance of being mv < 6, even at maximum, a thorough optical search for the putative star will almost certainly be needed. To aid in the immediate search for the supernova, Sky Publishing Corp. has established AstroAlert, a worldwide e-mail network of some 2,000 amateur and other small-telescope users. The announcement of a probable supernova will be made by the Supernova Neutrino Early Warning System and automatically forwarded to participants through AstroAlert. Details of the AstroAlert network will be described as well as protocols for validating the "guest" star and obtaining its precise position.

Optical Emission Lines from Warm Interstellar Clouds: A Decisive Test of the Decaying Neutrino Theory
Sciama, D. W.
Astrophysical Journal, 505, 1998/9/1, L35
Abstract - Recently developed instruments such as the Taurus Tunable Filter and the Wisconsin H alpha Mapper should be able to detect some or all of the optical emission lines H alpha , [O I] lambda 6300, [S II] lambda 6717, [N I] lambda 5200, and [N II] lambda 6584 from warm interstellar clouds, such as those observed by Spitzer & Fitzpatrick along the line of sight to the halo star HD 93521. The strengths of these lines should resolve the debate as to whether the free electrons, which Spitzer & Fitzpatrick held responsible for the observed excitation of C II in the clouds,

are located mainly in the skins of the clouds or in their interiors. If the free electrons are indeed mainly located in the cloud interiors, then the substantial electron density derived by Spitzer & Fitzpatrick and its constancy from cloud to cloud for the slow-moving clouds, when combined with their opacity to Lyman continuum radiation, lend strong support to the decaying neutrino theory for the ionization of the interstellar medium (Sciama). If the [O I] and [N I] lines are relatively strong but the [N II] line is weak, then this would lend further, decisive, support to this theory, since decay photons are unable to ionize N, although its ionization potential is only 0.9 eV greater than that of H.

Optimization of the Data Processing Algorithm of Searching for Neutrino-Gamma-Gravity Correlation
Gusev, A. V.; Milyukov, V. K.; Rudenko, V. N.; Vinogradov, M. P.
Second Edoardo Amaldi Conference on Gravitational Wave Experiments,
1998/1/1, 512
Abstract - Not Available

Oscillating Neutrinos from the Galactic Center
Crocker, Roland M.; Melia, Fulvio; Volkas, Raymond R.
Astrophysical Journal Supplement Series, 130, 2000/10/1, 339-350
Abstract - It has recently been demonstrated that the γ-ray emission spectrum of the EGRET-identified central Galactic source 2EG J1746-2852 can be well fitted by positing that these photons are generated by the decay of π⁰'s produced in p-p scattering at or near an energizing shock. Such scattering also produces charged pions which decay leptonically. The ratio of γ-rays to neutrinos generated by the central Galactic source can be accurately determined, and a well-defined and potentially measurable high-energy neutrino flux at Earth is unavoidable. An opportunity, therefore, to detect neutrino oscillations over an unprecedented scale is offered by this source. In this paper we assess the

prospects for such an observation with the generation of neutrino Cerenkov telescopes now in the planning stage. We determine that the next generation of detectors may well find an oscillation signature in the Galactic center (GC) signal.

Overview of Neutrino Oscillation Physics
Barger, V.
Particles, Strings and Cosmology,
2000/1/1, 377
Abstract - Not Available

Parametric Resonance of Neutrino Oscillations in Electromagnetic Wave
Dvornikov, M.; Studenikin, A.
New worlds in astroparticle physics,
2001/1/1, 126
Abstract - Not Available

Parity Violation in Neutrino Transport and the Origin of Pulsar Kicks: Erratum
Lai, Dong; Qian, Yong-Zhong
Astrophysical Journal, 501, 1998/7/1, L155
Abstract - In the Letter "Parity Violation in Neutrino Transport and the Origin of Pulsar Kicks" by D. Lai and Y.-Z. Qian (ApJ, 495, L103 [1998]), at the beginning of § 3, a citation of Horowitz & Li (1997; now updated to Phys. Rev. Lett., 80, 3694 [1998]) should be added to the end of the sentence "However, the cumulative effect due to multiple scatterings can enhance the asymmetry in neutrino emission." We have recently realized that in the bulk interior of the neutron star, where local thermodynamic equilibrium applies to a good approximation, detailed balance requires that there be no cumulative effect from multiple scatterings. Enhancement of neutrino emission asymmetry is obtained only after neutrinos thermally decouple from proto-neutron star matter. Therefore, our Letter overestimated the neutrino emission asymmetry and the resulting pulsar kicks.

Parity Violation in Neutrino Transport and the Origin of Pulsar Kicks
Lai, Dong; Qian, Yong-Zhong
Astrophysical Journal, 495, 1998/3/1, L103
Abstract - In proto-neutron stars with strong magnetic fields, the neutrino-

nucleon scattering/absorption cross sections depend on the direction of neutrino momentum with respect to the magnetic field axis, a manifestation of parity violation in weak interactions. We study the deleptonization and thermal cooling (via neutrino emission) of proto-neutron stars in the presence of such asymmetric neutrino opacities. Significant asymmetry in neutrino emission is obtained because of multiple neutrino-nucleon scatterings.

Particle physics, astrophysics and cosmology with forbidden neutrinos
Lindebaum, R. J.; Tupper, G. B.; Viollier, R. D.
Weak Interactions and Neutrinos, 2000/1/1, 200
Abstract - Not Available

Particles in the Bulk: A Higher-Dimensional Approach to Neutrino and Axion Phenomenology
Dienes, K. R.
Dark Matter in Astro- and Particle Physics, 2001/1/1, 234
Abstract - Not Available

Pauli's ghost: the conception and discovery of neutrinos
Riordan, Michael
Current aspects of neutrino physics, 2001/1/1, 1
Abstract - Contents: 1. Detecting neutrinos. 2. Massless neutrinos? 3. Two kinds of neutrinos. 4. The standard model. 5. The third family.

Perturbative quantum chromodynamics predictions for very high-energy atmospheric neutrinos and muons
Varieschi, Gabriele Umberto Ph.D.
Thesis, 2000/12/1, 7
Abstract - We compare the leading and next-to-leading order Quantum Chromodynamics predictions for the flux of atmospheric muons and neutrinos from decays of charmed particles. We then compute this flux for different Partonic Distribution Functions (PDF's) and different extrapolations of these at small partonic momentum fraction x. We find

that the predicted fluxes vary up to almost two orders of magnitude at the largest energies studied, depending on the chosen extrapolation of the PDF's. We show that the spectral index of the atmospheric leptonic fluxes depends linearly on the slope of the gluon distribution function at very small x. Finally, we analyze the uncertainties of our model and we consider the dependence of our simulation on the primary cosmic ray flux.

Phenomenological Analysis of CP-Violation in Neutrino Oscillation Experiments
Freund, M.
Proceedings of the Sixth SFB-375 Ringberg Workshop Astroteilchenphysik, 2000/2/1, 27
Abstract - Not Available

Phenomenology of atmospheric neutrinos
Lipari, P.; Lusignoli, M.
Weak Interactions and Neutrinos, 2000/1/1, 238
Abstract - Not Available

Plasma Acceleration of Photons (and Neutrinos)
Mendonca, J. T.
New Worlds in Astroparticle Physics, 1998/1/1, 417
Abstract - Not Available

Plasma instabilities driven by intense neutrino winds and anomalous heating in supernovae
Dawson, J. M.; Bingham, R.; Silva, L. O.
American Astronomical Society Meeting, 196, 2000/5/1
Abstract - In order to energize the outgoing shock and guarantee a successful supernova explosion, neutrinos must deposit about 1% of their energy in the plasma. Intense fluxes of photons and electrons can drive a whole class of plasma instabilities and deposit a significant amount of their free energy in the plasma. In the same way, ultra intense fluxes of neutrinos, released in type II supernovae and gamma-ray bursters, can drive collective longitudinal plasma modes, and transfer some of their free energy to the plasma. A wider class of plasma

instabilities can also be excited by neutrinos in particular those associated with the generation of electromagnetic fields (electroweak Weibel instability). We examine the electroweak generalization of the plasma instabilities driven by neutrinos. The linear growth rates for the different instability regimes are presented, and we show that the collective mechanisms are more important than the single particle collisional processes. Estimates for the pressure increase behind the stalled shock of SNe IIa are presented, including the contribution of the collective electromagnetic modes and the anomalous plasma heating due to damping of the electrostatic modes excited by the intense neutrino wind. Our results indicate that collective neutrino-plasma interactions can play an important role in the re-energization of the stalled shock, necessary for a successful SN explosion.

Ponderomotive Force of an Arbitrary Neutrino
Distribution in a Magnetized Plasma
Silva, L. O. et al.
New Worlds in Astroparticle Physics II,
1999/1/1, 220
Abstract - Not Available

Possible Relation between Relic Neutrinos and
the Highest Energy Cosmic Rays
Weiler, T.
Particles, Strings and Cosmology
(PASCOS 98), 1999/1/1, 297
Abstract - Not Available

Possible test for the suggestion that air showers
with E>10^20 eV are due to strongly
interacting neutrinos
Bordes, José; Chan, Hong-Mo; Faridani,
Jacqueline; Pfaudler, Jakov; Tsou, Sheung
Tsun
Astroparticle Physics, 8, 1998/2/1, 135-140
Abstract - The suggestion is made that air
showers with energies beyond the Greisen-
Zatsepin-Kuz'min spectral cutoff may have
primary vertices some 6 km lower in
height than those of proton initiated
showers with energies below the GZK
cutoff. This estimate is based on the
assumption that post-GZK showers are due
to neutrinos having acquired strong

interactions from generation-changing dual
gluon exchange as recently proposed.

Possible Universal Neutrino Interaction
Volkov, D. V.; Akulov, V. P.
LNP Vol. 509: Supersymmetry and
Quantum Field Theory, 1998/1/1, 383
Abstract - Not Available

Post-GZK Air Showers, FCNC, Strongly
Interacting Neutrinos and Duality
Bordes, J.; Hong-Mo, C.; Faridani, J.;
Pfaudler, J.; Tsun, T. S.
AIP Conf. Proc. 415: Beyond the Standard
Model. From Theory to Experiment,
1998/1/1, 328
Abstract - Not Available

Predictions of the Solar Neutrino Fluxes and
the Solar Gravity Mode Frequencies from
the Solar Sound Speed Profile
Turck-Chièze, S.; Brun, A. S.; Garcia, R.
A.
Structure and Dynamics of the Interior of
the Sun and Sun-like Stars SOHO
6/GONG 98 Workshop
Abstract - Recently, a lot of theoretical and
experimental efforts have been performed
in order to improve the knowledge of the
nuclear reaction rates, screening, opacity
calculations which are useful for a good
theoretical representation of the Sun. We
shall present these new works:
recompilation of all the cross sections
useful for the solar fusion (Aldelberger et
al 1998), measurements of the (3He,3He)
and (7Be, p) cross sections, new
calculations on screening enhancement,
introduction of more heavy elements in the
opacity coefficient calculations (Rogers
1998). The main progress will be discussed
through their effects on solar models,
neutrino and acoustic predictions (Brun,
Turck-Chièze and Morel 1998). A peculiar
attention will be devoted to the
confrontation with recent neutrino
measurements. One may notice that these
improvements play a signifant role at the
level of accuracy we are able to reach with
present seismology and that they are
extremely important for a reasonable
interpretation of what we learn from
helioseismology on the radiative region

and more precisely on the solar core. Considering the recent progress done by the ground networks and the SOHO satellite in helioseismology, the authors suggest new laboratory experiments on large lasers in order to disentangle different physical processes. Perspectives of what we prepare for the near future to better disentangle the neutrino puzzle will be illustrated.

Present Status of the Oto Cosmo Observatory and Neutrino Physics by Means of ELEGANT Detectors
Hayashi, K.; Ejiri, H.; Fushimi, K.; Hazama, R.; Kishimoto, T.; Komori, M.; Kume, K.; Kudomi, K.; Matsuoka, K.; Miyawaki, H.; Ohsumi, H.; Shiomi, S.; Takahisa, K.; Yoshida, S.
International Symposium on Origin of Matter and Evolution of Galaxies 97, 1998/1/1, 77
Abstract - Not Available

Primary Cosmic Ray Flux and Neutrino Oscillations
Honda, M.
New Era in Neutrino Physics , 1998/1/1, 139
Abstract - Not Available

Primary Cosmic-Ray Spectrum and the Intensity of Atmospheric Neutrinos
Gaisser, T. K.
New Era in Neutrino Physics , 1998/1/1, 145
Abstract - Not Available

Primordial Nucleosynthesis and Neutrino Cosmology
Kohri, K.; Kawasaki, M.; Sato, K.
Particle Cosmology, 1998/1/1, 109
Abstract - Not Available

Primordial Nucleosynthesis with Neutrino Degeneracy and Gravitational Constant Variation
Lee, Hyun Kyu
Recent Developments in Theoretical and Experimental General Relativity, Gravitation, and Relativistic Field Theories, 1999/1/1, 1462

Primordial Nucleosynthesis with Varying Gravitational Constant and Neutrino Degeneracy
Lee, H. K.
Pacific Conference on Gravitation and Cosmology, 1998/1/1, 138
Abstract - Not Available

Primordially produced helium-4 in the presence of neutrino oscillations
Kirilova, D. P.
Joint SOHO/ACE workshop "Solar and Galactic Composition", 2001/1/1, 405
Abstract - Not Available

Probing Early Universe Physics with Cosmic, Gamma-Ray, and Neutrino Astrophysics
Sigl, G.
Abstracts of the 19th Texas Symposium on Relativistic Astrophysics and Cosmology, held in Paris, France, Dec. 14-18, 1998. Eds.: J. Paul, T. Montmerle, and E. Aubourg (CEA Saclay)., 1998/12/1, 644
Abstract - New particle physics beyond the standard model is expected to play an important role in the early universe. Some of the possible resulting processes such as decaying topological defects or massive, long-lived dark matter particles, can contribute to the cosmic and gamma-radiation observed today. This data thus constrains and probes early universe physics. In addition, such processes could give rise to a relic ultra-high energy neutrino flux that could be detected with future neutrino telescopes. We discuss recent results and future prospects in this field.

Probing neutrino properties with the cosmic microwave background
Lopez, Robert Edwards Ph.D.
Thesis, 1999/1/1, 19
Abstract - Neutrinos that decay leave their imprint on the cosmic microwave background. We calculate the CMB anisotropy for the full range of decaying neutrino parameter space, and investigate the ability of future experiments like MAP and Planck to probe decaying neutrino physics. We adopt two approaches: distinguishing decaying neutrino models from fiducial ΛCDM, and

measuring neutrino parameters. With temperature data alone, MAP can distinguish stable neutrino models from ΛCDM, if the neutrino mass m_h>~ 2 eV. Adding neutrino parameters degrades the sensitivity to non-neutrino parameters; the relative amount of sensitivity degradation depends on the decaying neutrino model, but tends to decrease with increasing experimental sensitivity.

Probing the desert with ultra-energetic neutrinos from the sun and the earth
Faraggi, A. E.; Olive, K. A.; Pospelov, M.
Astroparticle Physics, 13, 2000/3/1, 31-43
Abstract - Realistic superstring models generically give rise to exotic matter states, which arise due to the "Wilson-line" breaking of the non-Abelian unifying gauge symmetry. Often such states are protected by a gauge or local discrete symmetry and therefore may be stable or meta-stable. We study the possibility of a flux of high energy neutrinos coming from the sun and the earth due to the annihilation of such exotic string states. We also discuss the expected flux for other heavy stable particles - like the gluino LSP. We comment that the detection of ultra-energetic neutrinos from the sun and the earth imposes model independent constraints on the high energy cutoff, as for example in the recently entertained TeV scale Kaluza-Klein theories. We therefore propose that improved experimental resolution of the energy of the muons in neutrino detectors together with their correlation with neutrinos from the sun and the center of the earth will serve as a probe of the desert in Gravity Unified Theories.

Probing the gravitational well: No supernova explosion in spherical symmetry with general relativistic Boltzmann neutrino transport
Liebendörfer, Matthias; Mezzacappa, Anthony; Thielemann, Friedrich-Karl; Messer, O. E.; Hix, W. Raphael; Bruenn, Stephen W.
Physical Review D, 63, 2001/5/1, 3004
Abstract - We report on the stellar core collapse, bounce, and postbounce evolution of a 13 M_{solar} star in a self-consistent general relativistic spherically symmetric simulation based on Boltzmann neutrino transport. We conclude that approximations to exact neutrino transport and the omission of general relativistic effects were not alone responsible for the failure of numerous preceding attempts to model supernova explosions in spherical symmetry. Compared to simulations in Newtonian gravity, the general relativistic simulation results in a smaller shock radius. We however argue that the higher neutrino luminosities and rms energies in the general relativistic case could lead to a larger supernova explosion energy.

Probing the interior of the Sun by measuring the solar neutrino flux
Bernabei, R.
Memorie della Societa Astronomica Italiana, 69, 1998/1/1, 555
Abstract - Not Available

Probing the Nature of Neutrinos at Pion Factories
Fazely, Ali R.
The Role of Neutrinos, Strings, Gravity, and Variable Cosmological Constant in Elementary Particle Physics, 2001/1/1, 113
Abstract - Not Available

Probing Unstable Massive Neutrinos with Current CMB Observations
Lopez, Robert E.
Abstracts of the 19th Texas Symposium on Relativistic Astrophysics and Cosmology, held in Paris, France, Dec. 14-18, 1998. Eds.: J. Paul, T. Montmerle, and E. Aubourg (CEA Saclay)., 1998/12/1, 575
Abstract - The pattern of anisotropies in the cosmic microwave background depends upon the masses and lifetimes of the three neutrino species. A neutrino species of mass greater than 10 eV with lifetime between 10^{13} and 10^{17} sec leaves a very distinct signature (due to the integrated Sachs-Wolfe effect): the anisotropies at large angles are predicted to be comparable to those on degree scales. Present data exclude such a possibility and hence this region of parameter space. For m_nu ~30 eV, tau ~10^{13} sec, we find

an interesting possibility: the integrated Sachs-Wolfe peak produced by the decaying neutrino in low-Omega models mimics the acoustic peak expected in an Omega = 1 model.

Production and Detection of Black Holes Using a Neutrino Array
Uehara, Y.
Progress of Theoretical Physics, 107, 2002/3/1, 621-624
Abstract - We consider the production of black holes through collisions high-energy cosmic neutrinos and nuclei contained in detectors. If the fundamental scale M$_*$ is O (TeV), as some higher-dimensional theories suggest, it may be possible to produce and observe about 10^2 -- 10^4 black holes per year using the ICECUBE detector.

Production of Light p-Process Isotopes in Neutrino-Irradiated Alpha-Rich Freezeouts
Swift, T. P.; Meyer, B. S.; The, L.-S.
American Astronomical Society Meeting, 197, 2000/12/1
Abstract - The origin of the light, neutron-capture bypassed (p-process) isotopes ^{92}Mo, ^{94}Mo, ^{96}Ru, and ^{98}Ru has long been a mystery. Sites that produce the majority of the p-process isotopes in correct solar proportions have long been known to underproduce the light species [1], thereby suggesting a different origin. The alpha-rich freezeout occurring near a nascent neutron star in Type II supernovae has been proposed [2,3,4]; however, only ^{92}Mo is strongly produced, and it is never the most overproduced isotope, as is required for its site of origin. We explore models of alpha-rich freezeouts that include simultaneous irradiation of the nuclei by the copious neutrinos emitted during the explosion. We find that neutrino-nucleus interactions significantly enhance production of the light p-process species both by affecting the electron-nucleon ratio during the nucleosynthesis and by increasing the charge of nuclei once nuclear quasi-equilibrium clusters have broken. This work was supported by the NSF Research Experiences for Undergraduates (REU) Site Program through grant AST 96169939 to Florida Tech and the Southeastern Association for Research in Astronomy (SARA). It was also supported by NSF grant AST 9819877 and NASA grant NAG5-4703 at Clemson University. References: [1] Woosley, S. E., and Howard, W. M. 1978, ApJS, 36, 285 [2] Woosley, S. E., and Hoffman, R. D. 1992, ApJ, 395, 202 [3] Fuller, G. M., and Meyer, B. S. 1995, ApJ, 453, 792 [4] Hoffman, R. D., Woosley, S. E., Fuller, G. M., and Meyer, B. S. 1996, ApJ, 460, 478

Production of neutrons, neutrinos and gamma-rays by a very fast pulsar in the Galactic Centre region
Bednarek, W.
Monthly Notices of the Royal Astronomical Society, 331, 2002/3/1, 483-487
Abstract - We consider the possibility that the excess of cosmic rays near ~10_{18} eV, reported by the AGASA and SUGAR groups from the direction of the Galactic Centre, is caused by a young, very fast pulsar in the high-density medium. The pulsar accelerates iron nuclei to energies ~10_{20} eV, as postulated by the Galactic models for the origin of the highest-energy cosmic rays. The iron nuclei, about 1yr after pulsar formation, leave the supernova envelope without energy losses and diffuse through the dense central region of the Galaxy. Some of them collide with the background matter creating neutrons (from disintegration of Fe), neutrinos and gamma-rays (in inelastic collisions). We suggest that neutrons produced at a specific time after the pulsar formation are responsible for the observed excess of cosmic rays at ~10_{18} eV. From normalization of the calculated neutron flux to the one observed in the cosmic ray excess, we predict the neutrino and gamma-ray fluxes. It has been found that the 1km$_2$ neutrino detector of the IceCube type should detect from a few up to several events per year from the Galactic Centre, depending on the parameters of the considered model. Moreover, future systems of Cherenkov telescopes (CANGAROO III, HESS, VERITAS) should be able to observe 1-10TeV

gamma-rays from the Galactic Centre if the pulsar was created inside a huge molecular cloud about 3-10×10_3 yr ago.

Progress Toward a Km-Scale Neutrino Detector in the Deep Ocean
Stokstad, R. G.
Particle and Nuclear Physics, 1998/1/1, 403
Abstract - Not Available

Progress towards OMNIS, the Observatory for Multi-flavor NeutrInos from Supernovae
Murphy, A. St. J.; Boyd, R. N.; Cline, D. B.; Colgate, S.; Fenyves, E.; Fuller, G. M.; Hencheck, M.; Hime, A.; Lee, K.; McLaughlin, G.; Meyer, B. S.; Mezzacappa, A.; Nieto, M. M.; Smith, P. F.; Vagins, M.; Vernon, W.; Zach, J. J.
American Astronomical Society Meeting, 195, 1999/12/1
Abstract - An observatory has been proposed which would allow the detection of a significant number of all flavors of neutrinos emitted from a Galactic supernova. The detection strategy utilizes the interactions of supernovae neutrinos with large volumes of materials from which neutrons would be spalled which would then be detected. This mechanism is especially sensitive to the mu- and tau-neutrino flavors, to which no other existing or planned detector has comparable sensitivity. The detailed measurements of neutrino fluxes that such an observatory would provide would allow significant probing of the core-collapse environment and mechanism, neutrino physics, and the possible formation of black-holes. Details of the current design work, which includes computer simulations and large volume liquid scintillator detector prototyping, will be presented.

Propagation of Cosmic Rays and Neutrinos Through Space
Yoshida, S.
AIP Conf. Proc. 433: Workshop on Observing Giant Cosmic Ray Air Showers From >10(20) eV Particles From Space, 1998/1/1, 217
Abstract - Not Available

Prospects for Present and Future Reactor Neutrino Experiments
Grassi, M.
The Identification of Dark Matter , 1999/1/1, 604
Abstract - Not Available

Prospects for Radio Detection of Extremely High Energy Cosmic Rays and Neutrinos in the Moon
Alvarez-Muñiz, J.; Zas, E.
AIP Conf. Proc. 579: Radio Detection of High Energy Particles, 2001/1/1, 128
Abstract - Not Available

Prospects of Hydroacoustic Detection of Ultra-High and Extremely High Energy Cosmic Neutrinos
Dedenko, L. G.; Karlik, Y. S.; Learned, J. G.; Svet, V. D.; Zheleznykh, I. M.
AIP Conf. Proc. 579: Radio Detection of High Energy Particles, 2001/1/1, 277
Abstract - Not Available

Prototype detector for ultrahigh energy neutrino detection
Klein, Joshua R.; Mann, Alfred K.
Astroparticle Physics, 10, 1999/5/1, 321-329
Abstract - Necessary technical experience is being gained from successful construction and deployment of current prototype detectors to search for UHE neutrinos in Antarctica, Lake Baikal in Russia, and the Mediterranean. The prototype detectors have also the important central purpose of determining whether or not UHE neutrinos do in fact exist in nature by observation of at least a few UHE neutrino-induced leptons with properties that are not consistent with expected backgrounds. We discuss here the criteria for a prototype detector to accomplish that purpose in a convincing way even if the UHE neutrino flux is substantially lower than predicted at present.

Pulsar Acceleration by Asymmetric Emission of Sterile Neutrinos
Nardi, Enrico; Zuluaga, Jorge I.
Astrophysical Journal, 549, 2001/3/1, 1076-1084

Abstract - A convincing explanation for the observed pulsar large peculiar velocities is still missing. We argue that any viable particle physics solution would most likely involve the resonant production of a noninteracting neutrino ν$_s$ of mass m$_{νs}$~20-50 keV. We propose a model where anisotropic magnetic field configurations strongly bias the resonant spin flavor precession of tau antineutrinos into ν$_s$. The asymmetric emission of ν $_s$ from the core can produce sizable natal kicks and account for recoil velocities of several hundred kilometers per second.

Pulsar Kicks from Asymmetric Neutrino Transport
Lai, D.; Arras, P.
AAS/High Energy Astrophysics Division, 31, 1999/4/1
Abstract - In magnetized proto-neutron stars, neutrino cross sections depend asymmetrically on the neutrino momenta due to parity violation in the weak interaction. As a consequence, neutrinos escaping from the collapsed core will develop a flux along the direction of the magnetic field in addition to the usual (spherical) diffusive flux. The neutron star will recoil opposite the direction of this asymmetric flux. We find that quite large magnetic fields B ~ 10(15}-10({16)) G are needed to produce recoil velocities of order a few hundred km s(-1) . Hence, the parity violation effect can only produce large space velocities for a relatively small percentage of the pulsar population with extremely large magnetic fields.

Pulsar Kicks from Neutrino Oscillations
Kusenko, A.; Segre, G.
Physical Review D, 59, 1999/3/1, 1302
Abstract - Neutrino oscillations can explain the observed motion of pulsars. We show that two different models of neutrino emission from a cooling neutron star are in good quantitative agreement and predict the same order of magnitude for the pulsar kick velocity, consistent with the data.

Pulsating pre-white dwarfs as laboratories for neutrino astrophysics
O'Brien, M. S.

ASP Conf. Ser. 169: 11th European Workshop on White Dwarfs, 1999/1/1, 78
Abstract - Not Available

Radiation of Angular Momentum by Neutrinos from Merged Binary Neutron Stars
Baumgarte, Thomas W.; Shapiro, Stuart L.
Astrophysical Journal, 504, 1998/9/1, 431
Abstract - We study neutrino emission from the remnant of an inspiraling binary neutron star following coalescence. The mass of the merged remnant is likely to exceed the stability limit of a cold, rotating neutron star. However, the angular momentum of the remnant may also approach or even exceed the Kerr limit, J/M2 = 1, so that total collapse may not be possible unless some angular momentum is dissipated. We find that neutrino emission is very inefficient in decreasing the angular momentum of these merged objects and may even lead to a small increase in J/M2. We illustrate these findings with a post-Newtonian, ellipsoidal model calculation. Simple arguments suggest that the remnant may form a bar mode instability on a timescale similar to or shorter than the neutrino emission timescale, in which case the evolution of the remnant will be dominated by the emission of gravitational waves.

Real time supernova neutrino burst detection with MACRO
Ambrosio, M.; Antolini, R.; Auriemma, G.; Baker, R.; Baldini, A.; Barbarino, G. C.; Barish, B. C.; Battistoni, G.; Bellotti, R.; Bemporad, C.; Bernardini, P.; Bilokon, H.; Bisi, V.; Bloise, C.; Bower, C.; Bussino, S.; Cafagna, F.; Calicchio, M.; Campana, D.; Carboni, M.; Castellano, M.; Cecchini, S.; Cei, F.; Celio, P.; Chiarella, V.; Corona, A.; Coutu, S.; de Benedictis, L.; de Cataldo, G.; Dekhissi, H.; de Marzo, C.; de Mitri, I.; de Vincenzi, M.; di Credico, A.; Erriquez, O.; Favuzzi, C.; Forti, C.; Fusco, P.; Giacomelli, G.; Giannini, G.; Giglietto, N.; Grassi, M.; Gray, L.; Grillo, A.; Guarino, F.; Guarnaccia, P.; Gustavino, C.; Habig, A.; Hanson, K.; Hawthorne, A.; Heinz, R.; Hong, J. T.; Iarocci, E.; Katsavounidis, E.; Kearns, E.; Kyriazopoulou, S.; Lamanna,

E.; Lane, C.; Levin, D. S.; Lipari, P.; Liu, R.; Longley, N. P.; Longo, M. J.; Ludlam, G.; Maaroufi, F.; Mancarella, G.; Mandrioli, G.; Manzoor, S.; Margiotta Neri, A.; Marini, A.; Martello, D.; Marzari-Chiesa, A.; Mazziotta, M. N.; Mazzotta, C.; Michael, D. G.; Mikheyev, S.; Miller, L.; Monacelli, P.; Montaruli, T.; Monteno, M.; Mufson, S.; Musser, J.; Nicoló, D.; Nolty, R.; Okada, C.; Orth, C.; Osteria, G.; Palamara, O.; Parlati, S.; Patera, V.; Patrizii, L.; Pazzi, R.; Peck, C. W.; Petrera, S.; Pistilli, P.; Popa, V.; Rainó, A.; Reynoldson, J.; Ronga, F.; Rubizzo, U.; Sanzgiri, A.; Satriano, C.; Satta, L.; Scapparone, E.; Scholberg, K.; Sciubba, A.; Serra-Lugaresi, P.; Severi, M.; Sioli, M.; Sitta, M.; Spinelli, P.; Spinetti, M.; Spurio, M.; Steinberg, R.; Stone, J. L.; Sulak, L. R.; Surdo, A.; Tarlé, G.; Togo, V.; Valente, V.; Walter, C. W.; Webb, R.
Astroparticle Physics, 8, 1998/2/1, 123-133
Abstract - The MACRO experiment has been running as a supernova neutrino detector since 1989 and is sensitive to the whole galaxy since the beginning of 1992. A galactic supernova would produce some hundreds of nu_e events in the detector. We describe our stellar gravitational collapse online monitors and alarm system, and present the results of a search for neutrino bursts from supernovae during a period of 1.5 yr.

Recent Development on Collective Neutrino Interactions
Bento, L.
New Worlds in Astroparticle Physics II, 1999/1/1, 225
Abstract - Not Available

Recent Results from Neutrino Oscillation Experiments at CERN
Fiorillo, G.
New Worlds in Astroparticle Physics II, 1999/1/1, 136
Abstract - Not Available

Recent Results in Neutrino Masses
Valle, J. W. F.
Particle and Nuclear Physics, 1998/1/1, 43
Abstract - Not Available

Recent Super-Kamiokande Result on Atmospheric Neutrino and K2K Long Baseline Neutrino Oscillation Experiment
Yanagisawa, C.
AIP Conf. Proc. 415: Beyond the Standard Model. From Theory to Experiment, 1998/1/1, 272
Abstract - Not Available

Recent Super-Kamiokande Result on Solar Neutrinos
Yanagisawa, C.
AIP Conf. Proc. 415: Beyond the Standard Model. From Theory to Experiment, 1998/1/1, 260
Abstract - Not Available

Reexamination of Standard Solar Model to the Solar Neutrino Problems
Fukasaku, K.; Fujita, T.
International Symposium on Origin of Matter and Evolution of Galaxies 97, 1998/1/1, 370
Abstract - Not Available

Registration of atmospheric neutrinos with the BAIKAL Neutrino Telescope NT-96
Balkanov, V. A.; Belolaptikov, I. A.; Bezrukov, L. B.; Budnev, N. M.; Chensky, A. G.; Danilchenko, I. A.; Djilkibaev, Zh.-A. M.; Domogatsky, G. V.; Doroshenko, A. A.; Fialkovsky, S. V.; Gaponenko, O. N.; Garus, A. A.; Gress, T. I.; Klabukov, A. M.; Klimov, A. I.; Klimushin, S. I.; Koshechkin, A. P.; Kulepov, V. F.; Kuzmichev, L. A.; Kuznetzov, V. E.; Laudinskaite, J. J.; Lovtzov, S. V.; Lubsandorzhiev, B. K.; Milenin, M. B.; Mirgazov, R. R.; Moseiko, N. I.; Netikov, V. A.; Osipova, E. A.; Panfilov, A. I.; Parfenov, Yu. V.; Pavlov, A. A.; Pliskovsky, E. N.; Pokhil, P. G.; Popova, E. G.; Rozanov, M. I.; Rubtzov, V. Yu.; Sokalski, I. A.; Spiering, Ch.; Streicher, O.; Tarashansky, B. A.; Thon, T.; Vasiljev, R. V.; Wischnewski, R.; Yashin, I. V.
Astroparticle Physics, 12, 1999/10/1, 75-86
Abstract - We present the first neutrino induced events observed with a deep underwater neutrino telescope. Data from 70 days effective life time of the BAIKAL prototype telescope NT-96 have been analyzed with two different methods. With

the standard track reconstruction method, 9 clear upward muon candidates have been identified, in good agreement with 8.7 events expected from Monte Carlo calculations for atmospheric neutrinos. The second analysis is tailored to muons coming from close to the opposite zenith. It yields 4 events, compared to 3.5 from Monte Carlo expectations. From this we derive a 90% upper flux limit of 1.1 .10^-13 cm^-2 sec^-1 for muons in excess of those expected from atmospheric neutrinos with zenith angle > 150 degrees and energy > 10 GeV.

Relativistic Effects on Neutrino Pair Annihilation above a Kerr Black Hole with the Accretion Disk
Asano, Katsuaki; Fukuyama, Takeshi
Astrophysical Journal, 546, 2001/1/1, 1019-1026
Abstract - Using idealized models of the accretion disk, we investigate the relativistic effects on the energy deposition rate via neutrino pair annihilation near the rotation axis of a Kerr black hole. Neutrinos are emitted from the accretion disk. The bending of neutrino trajectories and the redshift due to the disk rotation and gravitation are taken into consideration. The Kerr parameter, a, affects not only behavior of the neutrinos but also the inner radius of the accretion disk. When the deposition energy is mainly contributed by the neutrinos coming from the central part, the redshift effect becomes dominant as a becomes large, and the energy deposition rate is reduced compared with that neglecting the relativistic effects. On the other hand, for a small a, the bending effect becomes dominant and makes the energy increase by factor of 2 compared with that which neglects the relativistic effects. For the disk with a temperature gradient, the energy deposition rate for a small inner radius of the accretion disk is smaller than that estimated by neglecting the relativistic effects. The relativistic effects, especially for a large a, play a negative role in avoiding the baryon contamination problem in gamma-ray bursts.

Relic Neutrinos and Z-Resonance Mechanism for Highest Energy Cosmic Rays
Crooks, James L.; Dunn, James O.; Frampton, Paul H.
Astrophysical Journal, 546, 2001/1/1, L1-L3
Abstract - The origin of the highest energy cosmic rays remains elusive. The decay of a superheavy particle (X) into an ultraenergetic neutrino which scatters from a relic neutrino or antineutrino at the Z-resonance has attractive features. Given the necessary X mass of 10^{14}-10^{15} GeV, the required lifetime, 10^{15}-10^{16} yr, renders model building a serious challenge, but three logical possibilities are considered: (1) X is a Higgs scalar in SU(15) belonging to high-rank representation, leading to power-enhanced lifetime; (2) a global X quantum number has exponentially suppressed symmetry breaking by instantons; and (3) with additional space dimension(s), localization of X within the real-world brane leads to Gaussian decay suppression, the most efficient of the suppression mechanisms considered.

Relic Neutrinos, Monopoles, and Cosmic Rays above ~10^{20} eV
Weiler, T. J.
AIP Conf. Proc. 433: Workshop on Observing Giant Cosmic Ray Air Showers From >10(20) eV Particles From Space, 1998/1/1, 246
Abstract - Not Available

Resonance spin flavour precession and solar neutrinos
Pulido, J.; Akhmedov, E. K.
Astroparticle Physics, 13, 2000/5/1, 227-244
Abstract - We examine the prospects for the resonance spin flavour precession as a solution to the solar neutrino problem. We study seven different realistic solar magnetic field profiles and, by numerically integrating the evolution equations, perform a fit of the event rates for the three types of solar neutrino experiments (Ga, Cl and SuperKamiokande) and a fit of the energy spectrum of the recoil electrons in SuperKamiokande. A χ ^{2} analysis shows that the quality of the rate fits is

excellent for two of the field profiles and good for all others with χ ^{2}/d.o.f. always well below unity. Regarding the fits for the energy spectrum, their quality is better than that for the small mixing angle MSW solution of the solar neutrino problem, at the same level as that for the large mixing angle MSW solution but worse than that for the vacuum oscillations one. The experimental data on the spectrum are however largely uncertain especially in the high energy sector, so that it is too early yet to draw any clear conclusions on the likeliest type of particle physics solution to the solar neutrino problem.

Resonant active-sterile neutrino conversion and r-process nucleosynthesis in neutrino-heated supernova ejecta
Fetter, Jonathan Morton Ph.D.
Thesis, 2000/11/1, 5
Abstract - The neutrino-driven wind of a Type II supernova, is a likely site for the rapid neutron capture process (r-process) of heavy-element nucleosynthesis. Detailed models of the wind, however, have difficulty reproducing the solar system abundance pattern. In particular, models tend to find a ratio of neutrons to seed nuclei which is too low to account for the abundances of the heaviest elements, around A = 195. One way to increase the neutron-to-seed ratio is to increase the abundance of neutrons, or equivalently, to decrease the electron fraction in the wind. We study matter-enhanced neutrino transformation in the wind.

Results from Super-Kamiokande on Atmospheric Neutrinos
Stone, J.
AIP Conf. Proc. 444: Particle Physics and Cosmology, First Tropical Workshop, 1998/1/1, 3
Abstract - Not Available

Results from the NOMAD Neutrino Oscillation Experiment
Polesello, G.; The NOMAD Collaboration
The Identification of Dark Matter , 1999/1/1, 571

Abstract - Not Available

Results on 300 Days' Atmospheric Neutrino Data from Super-Kamiokande
Kaneyuki, K.
International Symposium on Origin of Matter and Evolution of Galaxies 97, 1998/1/1, 131
Abstract - Not Available

Results on neutrino oscillations from Super-Kamiokande
Wilkes, R. Jeffrey
Advances in Space Research, 26, 2001/1/1, 1813-1822
Abstract - Not Available

Role of group and phase velocity in high-energy neutrino observatories
Price, P. B.; Woschnagg, K.
Astroparticle Physics, 15, 2001/3/1, 97-100
Abstract - Kuzmichev recently showed that use of phase velocity rather than group velocity for Cherenkov light signals and pulses from calibration lasers in high-energy neutrino telescopes leads to errors in track reconstruction and distance measurement. We amplify on his remarks and show that errors for four cases of interest to AMANDA, IceCube, and radio Cherenkov detector are negligibly small.

Rotational and Related Periodicities in the Homestake and GALLEX Neutrino Data
Sturrock, P. A.; Walther, G.; Wheatland, M. S.
AAS/High Energy Astrophysics Division, 31, 1999/4/1
Abstract - If neutrinos have a sufficiently strong magnetic moment, the solar neutrino flux will be modulated by the Sun's internal magnetic field. We have spectrum-analyzed the Homestake data, looking for evidence of periodic modulation in the range 12.6 - 13.3 y(-1) due to structures in the radiative zone that has a sidereal rotation rate in the range 13.6 - 14.3 y(-1) . We find a peak at 12.88 y(-1) . The estimated probability of finding such a peak in the search band by chance is about 3%. We also find sidebands at 11.88, 12.88, 14.88 and 15.88 y(-1) , attributable to a seasonal modulation due to the tilt of

the solar axis. The estimated probability of this combination occurring by chance is about 0.2%. This work was supported in part by Air Force grant F49620-95-1-008 and NASA grants NAS 8-37334 and NAGW-5-4038.

Rotational Signature and Possible R-Mode Signature in the GALLEX Solar Neutrino Data
Sturrock, P. A.; Scargle, J. D.; Walther, G.; Wheatland, M. S.
Astrophysical Journal, 523, 1999/10/1, L177-L180
Abstract - Recent analysis of the Homestake data has yielded evidence that the solar neutrino flux varies in time-more specifically, that it exhibits a periodic variation that may be attributed to rotational modulation occurring deep in the solar interior, either in the tachocline or in the radiative zone. Here we present a spectral analysis of the GALLEX data that yields supporting evidence for this rotational modulation. The most prominent peak in the power spectrum occurs at the synodic frequency of 13.08 yr^{-1} (cycles per year) and is estimated to be significant at the 0.1% level. It appears that the most likely interpretation of this modulation is that the electron neutrinos have nonzero magnetic moment, so that they oscillate between left-hand (detectable) and right-hand (nondetectable) chiralities as they traverse the Sun's internal magnetic field. This oscillation could account for the neutrino deficit. The second strongest peak in the GALLEX spectrum has a period of 52 days, and this period occurs in other solar data as well. We suggest that this periodicity and also the Rieger 154 day periodicity, which shows up in many solar parameters and in the Homestake data, are due to r-mode oscillations.

Scattering of Ultrahigh Energy (UHE) Extragalactic Neutrinos onto Light Relic Neutrinos in Galactic HDM Halo Overcoming the GZK Cut off
Fargion, D.; Mele, B.
The Identification of Dark Matter , 1999/1/1, 597

Abstract - Not Available

Science and technology of Borexino: a real-time detector for low energy solar neutrinos Borexino Collaboration, G. Alimonti, C. Arpesella, H. Back, M. Balata, T. Beau, G. Bellini, J. Benziger, S. Bonetti, A. Brigatti, B. Caccianiga, L. Cadonati, F. Calaprice, G. Cecchet, M. Chen, A. DeBari, E. DeHaas, H. de Kerret, O. Donghi, M. Deutsch, F. Elisei, A. Etenko, F. von Feilitzsch, R. Fernholz, R. Ford, B. Freudiger, A. Garagiola, C. Galbiati, F. Gatti, S. Gazzana, M. Giammarchi, D. Giugni, A. Golubchikov, A. Goretti, C. Grieb, C. Hagner, T. Hagner, Salvo, C.; Scardaoni, R.; Schoenert, S.; Schuhbeck, K.; Seidel, H.; Shutt, T.; Simgen, H.; Sonnenschein, A.; Smirnov, O.; Sotnikov, A.; Skorokhvatov, M.; Sukhotin, S.; Tartaglia, R.; Testera, G.; Vogelaar, R.; Vitale, S.; Wojcik, M.; Zaimidoroga, O.; Zakharov, Y.
Astroparticle Physics, 16, 2002/1/1, 205-234
Abstract - Not Available

Science and technology of Borexino: a real-time detector for low energy solar neutrinos
Borexino Collaboration
Alimonti, G.; Arpesella, C.; Back, H.; Balata, M.; Beau, T.; Bellini, G.; Benziger, J.; Bonetti, S.; Brigatti, A.; Caccianiga, B.; Cadonati, L.; Calaprice, F.; Cecchet, G.; Chen, M.; DeBari, A.; DeHaas, E.; de Kerret, H.; Donghi, O.; Deutsch, M.; Elisei, F.; Etenko, A.; von Feilitzsch, F.; Fernholz, R.; Ford, R.; Freudiger, B.; Garagiola, A.; Galbiati, C.; Gatti, F.; Gazzana, S.; Giammarchi, M.; Giugni, D.; Golubchikov, A.; Goretti, A.; Grieb, C.; Hagner, C.; Hagner, T.; Hampel, W.; Harding, E.; Hartmann, F.; von Hentig, R.; Hess, H.; Heusser, G.; Ianni, A.; Inzani, P.; Kidner, S.; Kiko, J.; Kirsten, T.; Korga, G.; Korschinek, G.; Kryn, D.; Lagomarsino, V.; LaMarche, P.; Laubenstein, M.; Loeser, F.; Lombardi, P.; Magni, S.; Malvezzi, S.; Maniera, J.; Manno, I.; Manuzio, G.; Masetti, F.; Mazzucato, U.; Meroni, E.; Musico, P.; Neder, H.; Neff, M.; Nisi, S.; Oberauer, L.; Obolensky, M.;

Pallavicini, M.; Papp, L.; Perasso, L.; Pocar, A.; Raghavan, R.; Ranucci, G.; Rau, W.; Razeto, A.; Resconi, E.; Riedel, T.; Sabelnikov, A.; Saggese, P.; Salvo, C.; Scardaoni, R.; Schoenert, S.; Schuhbeck, K.; Seidel, H.; Shutt, T.; Simgen, H.; Sonnenschein, A.; Smirnov, O.; Sotnikov, A.; Skorokhvatov, M.; Sukhotin, S.; Tartaglia, R.; Testera, G.; Vogelaar, R.; Vitale, S.; Wojcik, M.; Zaimidoroga, O.; Zakharov, Y.
Astroparticle Physics, 16, 2002/1/1, 205-234
Abstract - Borexino, a real-time device for low energy neutrino spectroscopy is nearing completion of construction in the underground laboratories at Gran Sasso, Italy (LNGS). The experiment's goal is the direct measurement of the flux of ^{7}Be solar neutrinos of all flavors via neutrino-electron scattering in an ultra-pure scintillation liquid. Seeded by a series of innovations which were brought to fruition by large-scale operation of a 4-ton test detector at LNGS, a new technology has been developed for Borexino. It enables sub-MeV solar neutrino spectroscopy for the first time. This paper describes the design of Borexino, the various facilities essential to its operation, its spectroscopic and background suppression capabilities and a prognosis of the impact of its results towards resolving the solar neutrino problem. Borexino will also address several other frontier questions in particle physics, astrophysics and geophysics.

Scintillation Crystal Detector for Low Energy Neutrino and Astroparticle Physics
Wong, Henry; Li, Jin
ASP Conf. Ser. 241: The 7th Taipei Astrophysics Workshop on Cosmic Rays in the Universe, 2001/1/1, 123
Abstract - Not Available

Search for a possible space-time correlation between high energy neutrinos and gamma-ray bursts
Montaruli, T.; Ronga, F.
the MACRO collaboration
Astronomy and Astrophysics Supplement Series, 142, 2000/3/1, 447-449
Abstract - We look for space-time correlations between 2233 gamma-bursts in the Batse Catalogs and 894 upward-going muons produced by neutrino interactions in the rock below or inside MACRO. Considering a search cone of 10^{deg} around GRB directions and a time window of +/- 200 s we find 0 events to be compared to 0.035 expected background events due to atmospheric neutrinos. The corresponding upper limit (90% c.l.) is $0.87 \ 10^{-9} \ cm^{-2}$ upward-going muons per average burst. This article had been submitted as a contribution paper in the "Rome '98 gamma-Ray Burst workshop" and should have appeared in the A\&AS special issue for the workshop (A\&AS 1999, 138 (3)).

Search for EHE Neutrinos with the the HiRes Fly's Eye (Stage 1) detector.
Kieda, D.; The HIRES Collaboration et al.
Abstracts of the 19th Texas Symposium on Relativistic Astrophysics and Cosmology, held in Paris, France, Dec. 14-18, 1998. Eds.: J. Paul, T. Montmerle, and E. Aubourg (CEA Saclay)., 1998/12/1, 633
Abstract - The High Resoution Fly's Eye Stage 1 detector has been operational since May 1997. In the past year, the detector has collected an equivalent exposure to EHE neutrinos as the entire lifetime of the original Fly's Eye detector. We present the results from an initial search for EHE neutrinos detected by the Stage 1 HiRes detector.

Search for neutral heavy leptons in a high-energy neutrino beam
Vaitaitis, Arturas Genrikas Ph.D.
Thesis, 2000/10/1, 5
Abstract - A search for neutral heavy leptons (NHLs) has been performed using an instrumented decay channel at the NuTeV (E815) experiment at Fermilab. The decay channel was composed of helium bags interspersed with drift chambers, and was used in conjunction with the NuTeV neutrino detector to search for NHL decays. Our search was sensitive to isosinglet type neutral leptons with masses between 0.25 and 2.0 GeV. No evidence of NHLs was found. The experiment took place during the 1996-

1997 Fermilab fixed-target run. Collected data were examined for NHLs decaying into muonic final states (μμν, μ eν, μπ and μρ). This analysis places limits on the mixing of NHLs with standard light neutrinos at a level up to an order of magnitude more restrictive than previous search limits in this mass range.

Search for neutrino decay during the 1999 solar eclipse
Cecchini, S.; Giacomelli, G.; Hasegan, D.; Mandrioli, G.; Maris, O.; Patrizii, L.; Plaian, A.; Popa, V.; Stefanov, L.; Valeanu, V.
Abstract - The solar neutrino problem could arise from oscillation of one neutrino type into a second type. Neutrinos would have a mass and there could be the possibility of radiative neutrino decays. We discuss the search for neutrino decays during the 1999 solar eclipse: it involves the emitted visible photons, while neutrinos travel from the Moon to the Earth. The concept and the main characteristics of the NOTTE experiment are presented.

Search for Neutrino Mass and Oscillations with a Powerful Supernova Detector Array
Cline, D. B.
The Identification of Dark Matter , 1999/1/1, 668
Abstract - Not Available

Search for supernova neutrino bursts with the AMANDA detector
Ahrens, J.; Bai, X.; Barouch, G.; Barwick, S. W.; Bay, R. C.; Becka, T.; Becker, K.-H.; Bertrand, D.; Biron, A.; Booth, J.; Botner, O.; Bouchta, A.; Boyce, M. M.; Carius, S.; Chen, A.; Chirkin, D.; Conrad, J.; Cooley, J.; Costa, C. G. S.; Cowen, D. F.; Dalberg, E.; DeYoung, T.; Desiati, P.; Dewulf, J.-P.; Doksus, P.; Edsjö, J.; Ekström, P.; Feser, T.; Gaug, M.; Goldschmidt, A.; Hallgren, A.; Halzen, F.; Hanson, K.; Hardtke, R.; Hellwig, M.; Heukenkamp, H.; Hill, G. C.; Hulth, P. O.; Hundertmark, S.; Jacobsen, J.; Karle, A.; Kim, J.; Koci, B.; Köpke, L.; Kowalski, M.; Lamoureux, J. I.; Leich, H.; Leuthold, M.; Lindahl, P.; Liubarsky, I.; Loaiza, P.; Lowder, D. M.; Madsen, J.; Marciniewski, P.; Matis, H. S.; Miller, T. C.; Minaeva, Y.; Miocinovic, P.; Mock, P. C.; Morse, R.; Neunhöffer, T.; Niessen, P.; Nygren, D. R.; Ogelman, H.; Pérez de los Heros, C.; Porrata, R.; Price, P. B.; Rawlins, K.; Reed, C.; Rhode, W.; Richter, S.; Rodríguez Martino, J.; Romenesko, P.; Ross, D.; Sander, H.-G.; Schmidt, T.; Schneider, D.; Schwarz, R.; Silvestri, A.; Solarz, M.; Spiczak, G. M.; Spiering, C.; Starinsky, N.; Steele, D.; Steffen, P.; Stokstad, R. G.; Streicher, O.; Sudhoff, P.; Taboada, I.; Thollander, L.; Thon, T.; Tilav, S.; Vander Donckt, M.; Walck, C.; Weinheimer, C.; Wiebusch, C. H.; Wischnewski, R.; Wissing, H.; Woschnagg, K.; Wu, W.; Yodh, G.; Young, S.
Astroparticle Physics, 16, 2002/2/1, 345-359
Abstract - The core collapse of a massive star in the Milky Way will produce a neutrino burst, intense enough to be detected by existing underground detectors. The AMANDA neutrino telescope located deep in the South Pole ice can detect MeV neutrinos by a collective rate increase in all photo-multipliers on top of dark noise. The main source of light comes from positrons produced in the CC reaction of anti-electron neutrinos on free protons ν$^{-}_{e}$+p-->e^{+}+n. This paper describes the first supernova search performed on the full sets of data taken during 1997 and 1998 (215 days of live time) with 302 of the detector's optical modules. No candidate events resulted from this search. The performance of the detector is calculated, yielding a 70/% coverage of the galaxy with one background fake per year with 90/% efficiency for the detector configuration under study. An upper limit at the 90/% c.l. on the rate of stellar collapses in the Milky Way is derived, yielding 4.3 events per year. A trigger algorithm is presented and its performance estimated. Possible improvements of the detector hardware are reviewed.

Search for the Antineutrino Rest Mass in the
Tritium Beta Decay
Lobashev, V. M.
Particle and Nuclear Physics, 1998/1/1,
337
Abstract - Not Available

Search for WIMPs at neutrino telescopes
Hulth, Per Olof
Abstracts of the 19th Texas Symposium on
Relativistic Astrophysics and Cosmology,
held in Paris, France, Dec. 14-18, 1998.
Eds.: J. Paul, T. Montmerle, and E.
Aubourg (CEA Saclay)., 1998/12/1, 272
Abstract - The WIMPs (weakly interacting
massive particles, eg neutralinos) might
explain part of the dark matter problem in
the Universe. Several experiments are
searching for neutrinos coming from
WIMP annihilations in the centre of the
Earth and in the centre of the Sun. The
status of these searches will be reviewed.

Searches for neutrino mass in the NuTeV
experiment
Shaevitz, M. H. et al.
Weak Interactions and Neutrinos,
2000/1/1, 400
Abstract - Not Available

Searches for neutrino oscillations: LSND
Garvey, G. T.; LSND Collaboration
Lepton and Baryon Number Violation in
Particle Physics, neral Astrophysics and
Cosmology, 1999/1/1, 227
Abstract - Not Available

Searching for Low Energy Electron Anti-
Neutrinos from the Sun
Semikoz, V. B.; Pastor, S.; Valle, J. W. F.
AIP Conf. Proc. 415: Beyond the Standard
Model. From Theory to Experiment,
1998/1/1, 306
Abstract - Not Available

Searching for the MSW Solar Neutrino
Oscillation Enhancement Using the Earth
Gelb, J. M.; Kwong, W.; Rosen, S. P.
American Astronomical Society Meeting,
192, 1998/5/1
Abstract - We point out that the length
scale associated with the MSW effect is the
radius of the Earth. Therefore to verify
matter enhancement of neutrino
oscillations, it will be necessary to study
neutrinos passing through the Earth. For
the parameters of MSW solutions to the
solar neutrino problem, the only detectable
effects occur in a narrow band of energies
from 5 to 10 MeV. We propose that serious
consideration be given to mounting an
experiment at a location within 9.5 degrees
of the equator.

Seasonal Variations in Solar High-Energy
Neutrino Flux and Their Probable Source
Rivin, Yu. R.; Obridko, V. N.
Solar System Research, 34, 2000/11/1,
501-508
Abstract - Cyclic variations of the solar
neutrino flux (Homestake detector data)
have been analyzed both from season to
season and within different seasons and
were compared with the corresponding
variations of the large-scale deep-layer
solar magnetic field. The analysis revealed
a seasonal variation of the flux in the last
twenty years with extremes at equinox
epochs. The mechanism of this variation
can be due to the asymmetry in magnitudes
or to the twisting of the large-scale
magnetic fields in the southern and
northern hemispheres of the Sun in the flux
modulation region.

Secondary decays in atmospheric charm
contributions to the flux of muons and
muon neutrinos
Pasquali, L.; Reno, M. H.; Sarcevic, I.
Astroparticle Physics, 9, 1998/10/1, 193-
202
Abstract - We present a calculation of the
fluxes of muons and muon neutrinos from
the decays of pions and kaons that are
themselves the decay products of charmed
particles produced in the atmosphere by
cosmic ray-air collisions. Using the
perturbative cross section for charm
production, these lepton fluxes are two to
three orders of magnitude smaller than the
fluxes from the decays of pions and kaons
directly produced in cosmic ray-air
collisions. Intrinsic charm models do not
significantly alter our conclusions for
E<10^3 GeV. Models with a charm
cross section enhanced in the region above

an incident cosmic ray energy of 10^3 GeV give enhanced lepton fluxes above 10^4 GeV, a region in which prompt muons dominate over the secondary decay contributions.

Seismic constraints on neutrino oscillation parameters
Goswami, Srubabati; Kar, Kamales; Antia, H. M.; Chitre, S. M.
Proceedings of the SOHO 10/GONG 2000 Workshop: Helio- and asteroseismology at the dawn of the millennium, 2-6 October 2000, Santa Cruz de Tenerife, Tenerife, Spain. Edited by A. Wilson, Scientific coordination by P. L. Pallé. ESA SP-464, Noordwijk: ESA Publications Division, ISBN 92-9092-697-X, 2001, p. 519 - 522, 464, 2001/1/1, 519
Abstract - The neutrino fluxes calculated using a seismically inferred solar model are compared with measured fluxes from the three solar neutrino experiments. Treating the neutrino fluxes from seismic model as theoretical predictions, the latest solar neutrino data is analyzed assuming vacuum oscillation of neutrinos. The best-fit values of the neutrino mixing angle and mass squared difference are found and the allowed regions are determined.

Seismic solar models and the neutrino problem
Takata, M.; Shibahashi, H.
IAU Symp. 185: New Eyes to See Inside the Sun and Stars, 185, 1998/1/1, 21
Abstract - Not Available.

Semi-analytic approximations for production of atmospheric muons and neutrinos
Gaisser, Thomas K.
Astroparticle Physics, 16, 2002/1/1, 285-294
Abstract - Simple approximations for fluxes of atmospheric muons and muon neutrinos are developed which display explicitly how the fluxes depend on primary cosmic-ray energy and on features of pion production. For energies of approximately 10 GeV and above the results are sufficiently accurate to calculate response functions and to use for estimates of systematic uncertainties.

Sensitivity of an underwater acoustic array to ultra-high energy neutrinos
Lehtinen, Nikolai G.; Adam, Shaffique; Gratta, Giorgio; Berger, Thomas K.; Buckingham, Michael J.
Astroparticle Physics, 17, 2002/6/1, 279-292
Abstract - We investigate the possibility of searching for ultra high energy neutrinos in cosmic rays using acoustic techniques in ocean water. The type of information provided by the acoustic detection is complementary to that of other techniques, and the filtering effect of the atmosphere, imposed by the fact that detection only happens if a shower fully develops in water, would provide a clear neutrino identification. We find that it may be possible to implement this technique with very limited resources using existing high frequency underwater hydrophone arrays. We review the expected acoustic signals produced by neutrino-induced showers in water and develop an optimal filtering algorithm able to suppress statistical noise.

Sensitivity of Borexino to Seasonal Variations of the Solar Neutrino Flux
Maneira, J. C.
New Worlds in Astroparticle Physics II, 1999/1/1, 131
Abstract - Not Available

Shadows of Relic Neutrino Masses and Spectra on Highest Energy GZK Cosmic Rays
Fargion, D. et al.
Dark Matter in Astro- and Particle Physics, 2001/1/1, 455
Abstract - Not Available

Simulation of Neutrino Transport by Large-Scale Convective Instability in a Proto-Neutron Star
Suslin, V. M.; Ustyugov, S. D.; Chechetkin, V. M.; Churkina, G. P.
Astronomy Reports, 45, 2001/3/1, 241-247
Abstract - Not Available

Simulations of Core Collapse Supernovae in One and Two Dimensions Using Multigroup Neutrino Transport
Mezzacappa, A.
Eighteenth Texas Symposium on

Relativistic Astrophysics, 1998/1/1, 714
Abstract - Not Available

Sneutrino physics with lepton number violation
Kolb, St.; Klapdor-Kleingrothaus, H. V.;
Hirsch, M.; Panella, O.
Lepton and Baryon Number Violation in
Particle Physics, neral Astrophysics and
Cosmology, 1999/1/1, 621
Abstract - Not Available

SNEWS: A Neutrino Early Warning System for
Galactic SN II
Habig, A.; SNEWS Collaboration
American Astronomical Society Meeting,
195, 1999/12/1
Abstract - The detection of neutrinos from
SN1987A confirmed the core-collapse
nature of SN II, but the analysis was done
after the optical discovery. The current
generation of neutrino experiments are
both much larger and actively looking for
SN neutrinos in real time. Since neutrinos
escape a new SN promptly while the first
photons are not produced until hours later
at the photospheric shock breakout, these
experiments can provide an early warning
of a galactic SN II. Such an advance notice
of a coming nearby supernova would allow
observations at all wavelengths from the
earliest possible times. A coincidence
network between neutrino experiments has
been established to mimimize response
time, eliminate experimental false alarms,
and possibly provide some pointing to the
impending event from neutrino wave-front
timing.

SNO Observables from the Magnetic Moment
Solution to the Solar Neutrino Problem
Pulido, J.
New worlds in astroparticle physics,
2001/1/1, 121
Abstract - Not Available

Solar and atmospheric neutrino oscillations --
Super-Kamiokande results
Suzuki, Y.; The Super-KAMIOKANDE
Collaboration
Weak Interactions and Neutrinos,
2000/1/1, 277
Abstract - Not Available

Solar and Atmospheric Neutrino Oscillations -
Super-Kamiokande Results
Suzuki, Y.
New Worlds in Astroparticle Physics II,
1999/1/1, 3-5
Abstract - Not Available

Solar and Atmospheric Neutrino Results from
Super-Kamiokande
Conner, Z.
25th International Cosmic Ray Conference,
Volume 8, 8, 1998/1/1, 291
Abstract - Not Available

Solar and Atmospheric Neutrinos
Kajita, Takaaki
Particles, Strings and Cosmology,
2000/1/1, 386
Abstract - Not Available

Solar and Supernova Constraints of
Cosmologically Interesting Neutrinos
Haxton, W. C.
NATO ASIC Proc. 511: Current Topics in
Astrofundamental Physics: Primordial
Cosmology, 1998/1/1, 703
Abstract - Not Available

Solar Magnetic Field from Solar Neutrino
Experiments
Pulido, J.
New Worlds in Astroparticle Physics II,
1999/1/1, 142
Abstract - Not Available

Solar Models Based on Helioseismology and
the Solar Neutrino Problem
Takata, Masao; Shibahashi, Hiromoto
Astrophysical Journal, 504, 1998/9/1, 1035
Abstract - We have determined the sound-
speed profile in the Sun by carrying out an
asymptotic inversion of the helioseismic
data from the Low-Degree (l) Oscillation
Experiment (LOWL), the Global
Oscillation Network Group (GONG),
VIRGO on SOHO, the High-l
Helioseismometer (HLH), and
observations made at the South Pole. We
then deduce the density, pressure,
temperature, and elemental composition
profiles in the solar radiative interior by
solving the basic equations governing the
stellar structure, with the imposition of the

determined sound-speed profile and with a constraint on the depth of the convection zone obtained from helioseismic analysis and the ratio of the metal abundance to the hydrogen abundance at the photosphere. With the exception of the treatment of elements relevant to nuclear reactions, we assume that Z is homogeneous. The chemical composition profiles of hydrogen and helium are then obtained as a part of the solutions. Using the resulting seismic model, we estimate the neutrino fluxes and the neutrino capture rates for the chlorine, gallium, and water Cerenkov experiments.

Solar Models with Helioseismic Constraints and the Solar Neutrino Problem
Watanabe, Satoru; Shibahashi, Hiromoto
Publications of the Astronomical Society of Japan, 53, 2001/6/1, 565-575
Abstract - Imposing a constraint of the sound-speed profile determined from helioseismology and updating the microphysics, we have revised our seismic solar model, constructed with the assumption of a homogeneous metal abundance distribution, and have shown that the theoretically expected neutrino fluxes are still significantly more than the observations. With the same sound-speed profile constraint, we also constructed solar models with low metal abundance in the core, and evaluated the neutrino fluxes of these models to see if nonstandard solar models with a low metal core can solve the solar neutrino problem.

Solar models: constraints from helioseismology and neutrino production
degl'Innocenti, S.; Castellani, V.; Dziembowski, W. A.; Fiorentini, G.; Ricci, B.
Memorie della Societa Astronomica Italiana, 69, 1998/1/1, 539
Abstract - Not Available

Solar Models: Current Epoch and Time Dependences, Neutrinos, and Helioseismological Properties ; Bahcall, John N.; Pinsonneault, M. H.; Basu, Sarbani
Astrophysical Journal, 555, 2001/7/1, 990-1012

Abstract - We calculate accurate solar models and report the detailed time dependences of important solar quantities. We use helioseismology to constrain the luminosity evolution of the Sun and report the discovery of semiconvection in evolved solar models that include diffusion. In addition, we compare the computed sound speeds with the results of p-mode observations by BiSON, GOLF, GONG, LOWL, and MDI instruments. We contrast the neutrino predictions from a set of eight standard-like solar models and four deviant (or deficient) solar models with the results of solar neutrino experiments. For solar neutrino and helioseismological applications, we present present-epoch numerical tabulations of characteristics of the standard solar model as a function of solar radius, including the principal physical and composition variables, sound speeds, neutrino fluxes, and functions needed for calculating solar neutrino oscillations.

Solar neutrino detection: recent results and future perspectives
Bellotti, E.; Ferrari, N.
Memorie della Societa Astronomica Italiana, 72, 2001/1/1, 500
Abstract - The measurement of the solar neutrino energy spectrum offers a unique opportunity to investigate the physics of the solar interior, and also to test non standard neutrino properties such as flavour mixing and magnetic moment. We give here a review of the present experimental results, and the perspectives for the future.

Solar Neutrino Emission Deduced from a Seismic Model
Turck-Chièze, S.; Couvidat, S.; Kosovichev, A. G.; Gabriel, A. H.; Berthomieu, G.; Brun, A. S.; Christensen-Dalsgaard, J.; García, R. A.; Gough, D. O.; Provost, J.; Roca-Cortes, T.; Roxburgh, I. W.; Ulrich, R. K.
Astrophysical Journal, 555, 2001/7/1, L69-L73
Abstract - Three helioseismic instruments on the Solar and Heliospheric Observatory have observed the Sun almost continuously

since early 1996. This has led to detailed study of the biases induced by the instruments that measure intensity or Doppler velocity variation. Photospheric turbulence hardly influences the tiny signature of conditions in the energy-generating core in the low-order modes, which are therefore very informative. We use sound-speed and density profiles inferred from GOLF and MDI data including these modes, together with recent improvements to stellar model computations, to build a spherically symmetric seismically adjusted model in agreement with the observations. The model is in hydrostatic and thermal balance and produces the present observed luminosity. In constructing the model, we adopt the best physics available, although we adjust some fundamental ingredients, well within the commonly estimated errors, such as the p-p reaction rate (+1%) and the heavy-element abundance (+3.5%); we also examine the sensitivity of the density profile to the nuclear reaction rates. Then, we deduce the corresponding emitted neutrino fluxes and consequently demonstrate that it is unlikely that the deficit of the neutrino fluxes measured on Earth can be explained by a spherically symmetric classical model without neutrino flavor transitions. Finally, we discuss the limitations of our results and future developments.

Solar neutrino experiment: current status and perspectives
Bellotti, E.
Weak Interactions and Neutrinos, 2000/1/1, 291
Abstract - Not Available

Solar neutrino in relation to solar wind particles
Basu, D.
Solar Physics, 184, 1999/1/1, 153-156
Abstract - A relationship was found earlier (Basu, 1982, 1992) between the solar neutrino flux and the flux of solar wind particles received on the Earth. However, the data used in these analyses have recently been revised and extended. This prompted us to re-examine the relationship using the new updated solar neutrino data

base. The present analysis confirms the earlier findings and establishes that the two quantities are related at statistically significant levels.

Solar Neutrino Observation in Super-Kamiokande
Nakahata, M.; The Super-KAMIOKANDE Collaboration
New Era in Neutrino Physics , 1998/1/1, 93
Abstract - Not Available

Solar Neutrino Results from Super-Kamiokande
Sullivan, G.
Super-Kamiokande Collaboration
AAS/High Energy Astrophysics Division, 32, 2000/10/1
Abstract - I report on recent results using solar neutrinos from the Super-Kamiokande detector. The solar neutrino flux, energy spectrum, day-night flux variation and seasonal flux variation are given. The implications of these measurements on the solar neutrino problem will be presented in the context of neutrino oscillations. I gratefully acknowledge the support of the National Science Foundation.

Solar Neutrino Spectroscopy with Borexino and Recent Results from the Ctf Experiment
von Feilitzsch, F.
BOREXINO Collaboration
Particle and Nuclear Physics, 1998/1/1, 123
Abstract - Not Available

Solar Neutrinos
Bahcall, J. N.
New Worlds in Astroparticle Physics II, 1999/1/1, 28
Abstract - Not Available

Solar Neutrinos
Suzuki, Y.
Solar Composition and Its Evolution -- From Core to Corona, 1998/1/1, 91
Abstract - Not Available

Solar Neutrinos
Suzuki, Y.

Space Science Reviews, 85, 1998/8/1, 91-104
Abstract - The current status of solar neutrino experiments is reviewed. All the experimental measurements show deficits of solar neutrinos. Non monotonic suppression indicates that the problem may naturally be explained by neutrino oscillations, but not by modifying solar models. A new experiment shows very promising results. We hope that a definite answer to the question of whether solar neutrinos are oscillating will be obtained in the very near future.

Solar Neutrinos as a Possible Diagnostic of the Solar Magnetic Field
Sturrock, P. A.
ASP Conf. Ser. 206: High Energy Solar Physics Workshop - Anticipating Hess!, 2000/1/1, 83
Abstract - Not Available

Solar Neutrinos as Highlight of Astroparticle Physics
Berezinsky, V.
25th International Cosmic Ray Conference, Volume 8, 8, 1998/1/1, 59
Abstract - Not Available

Solar neutrinos, atmospheric neutrinos and proton decays in Super-Kamiokande and the Kam-LAND project
Suzuki, A.
Lepton and Baryon Number Violation in Particle Physics, neral Astrophysics and Cosmology, 1999/1/1, 189
Abstract - Not Available

Solar Neutrinos: An Overview
Bahcall, J. N.
Confluence , of Cosmology, Massive Neutrinos, Elementary Particles, and Gravitation, 1999/1/1, 37
Abstract - Not Available

Solar Neutrinos: Where We Are
Bahcall, J.
Eighteenth Texas Symposium on Relativistic Astrophysics, 1998/1/1, 99
Abstract - Not Available

Solar Neutrinos: Where We Are
Bahcall, J.
Particles, Strings and Cosmology (PASCOS 98), 1999/1/1, 2-29
Abstract - Not Available

Solar Neutrinos: Where We are and What is Next?
Fiorentini, G.; Ricci, B.
AIP Conf. Proc. 415: Beyond the Standard Model. From Theory to Experiment, 1998/1/1, 241
Abstract - Not Available

Solar Structure After Neutrinos and Helioseismology (CD-ROM Directory: contribs/gough)
Gough, D. O.
ASP Conf. Ser. 223: 11th Cambridge Workshop on Cool Stars, Stellar Systems and the Sun, 11, 2001/1/1, 83
Abstract - Not Available

Solar-Neutrino Problem Solved
MacRobert, Alan M.; Tytell, David ; Sky and Telescope, 102, 2001/9/1, 18
Abstract - Not Available

Some Particle Physics Aspects of Neutrinoless Double Beta Decay
Hirsch, M.; Klapdor-Kleingrothaus, H. V.
Particle and Nuclear Physics, 1998/1/1, 323
Abstract - Not Available

Some Remarks on the Neutrino Oscillation Phase in a Gravitational Field (Letter)
Pereira, J. G.; Zhang, C. M.
General Relativity and Gravitation, 32, 2000/8/1, 1633-1637
Abstract - Not Available

Some Topics in Cosmology and Neutrino Physics
Song, Jeonghyeon Ph.D.
Thesis, 1998/7/1, 18
Abstract - This dissertation is devoted to the study of phenomenology in neutrino physics and cosmology: gravitational effects on the neutrino oscillation, finite temperature effects on the neutrino decoupling in the early Universe, and a scale-dependent cosmology. The main

results are: (1) The propagation and oscillation of neutrinos in a gravitational field are studied by calculating a covariant quantum-mechanical phase in a curved spacetime. Specifically, the cases of the radial propagation and the non-radial propagation in the Schwarzschild metric are considered. A possible application to gravitational lensing of neutrinos is also suggested. (2) Leading finite temperature effects on the neutrino decoupling temperature in the early Universe are studied. By modifying the dispersion relations and the phase space distribution functions due to the presence of the particles in the heat bath at temperature around MeV we have shown that the finite temperature effects increase the neutrino decoupling temperature by 4.4%, the largest contribution coming from the modification of the phase space distribution. (3) We propose a scale-dependent cosmology with a stress-energy tensor of viscous fluid, in which the Robertson-Walker metric and Einstein field equations are generalized in such a way that Ω $_0/H_0$ and the age of the Universe all become scale-dependent. Its implications on observations and possible modifications of the standard Friedmann cosmology are discussed, which suggest a picture of a Universe inside a highly dense and rapidly expanding shell with the underdense center. In this model, the linear Hubble diagram for nearby objects (z/lll) remains valid. For large z, we present some numerical results of the luminosity distance and the mass density against the redshift. It is shown that the Hubble diagram in this model for a locally open Universe resembles to that of the flat Friedmann cosmology and Ω $_0$ ^{obs} is an increasing function of the redshift.

Spectrum of solar neutrinos above 6.5 MeV
Sanford, Robert Ellis, Jr. Ph.D.
Thesis, 1999/12/1, 3
Abstract - The spectrum of recoil electrons from solar neutrino scattering above 6.5 MeV has been measured using the first 504 days of Super-Kamiokande detector data. The scattering rate is found to be 13.56 +/- 0.42(stat.) +/- 0.29(syst.)

events/day/22.5kton, which is a factor of 0.474 +/- 0.015(stat.) +/- 0.010(syst.) of the expected rate. The measured spectrum and the expected spectra from ^{8}B and HeP Neutrino scattering are compared using a χ ^{2} minimization process to find the best-fit match between the measured neutrino rates and a linear combination of the expected rates from ^{8}B and HeP neutrinos. Using only the ^{8}B expected spectrum the fitting procedure results in a best-fit scaling factor of 0.479; that is, the ^{8}B expected spectrum best matches the measured spectrum if it is scaled by 0.479.

Spherical collapse of supermassive stars: Neutrino emission and gamma-ray bursts
Linke, F.; Font, J. A.; Janka, H.-T.; Müller, E.; Papadopoulos, P.
Astronomy and Astrophysics, 376, 2001/9/1, 568-579
Abstract - We present the results of numerical simulations of the spherically symmetric gravitational collapse of supermassive stars (SMS). The collapse is studied using a general relativistic hydrodynamics code. The coupled system of Einstein and fluid equations is solved employing coordinates adapted to a foliation of the spacetime by means of outgoing null hypersurfaces. The code contains an equation of state which includes effects due to radiation, electrons and baryons, and detailed microphysics to account for electron-positron pairs. In addition energy losses by thermal neutrino emission are included. We are able to follow the collapse of SMS from the onset of instability up to the point of black hole formation. Several SMS with masses in the range 5x 10^5 M_{sun}-10^9 M_{sun} are simulated. In all models an apparent horizon forms initially, enclosing the innermost 25% of the stellar mass. From the computed neutrino luminosities, estimates of the energy deposition by nu bar nu-annihilation are obtained. Only a small fraction of this energy is deposited near the surface of the star, where, as proposed recently by Fuller & Shi (\cite{Fuller98}), it could cause the ultrarelativistic flow believed to be responsible for gamma -ray bursts. Our

simulations show that for collapsing SMS with masses larger than 5x 10^5 M_{sun} the energy deposition is at least two orders of magnitude too small to explain the energetics of observed long-duration bursts at cosmological redshifts. In addition, in the absence of rotational effects the energy is deposited in a region containing most of the stellar mass. Therefore relativistic ejection of matter is impossible.

Spherically Symmetric Core Collapse Supernova Simulations With Boltzmann Neutrino Transport
Messer, O. E. B.
American Astronomical Society Meeting, 199, 2001/12/1
Abstract - I will describe the results of several spherically symmetric core collapse supernova simulations performed with AGILE-BOLTZTRAN, a state-of-the-art radiation hydrodynamics code incorporating Boltzmann neutrino transport. Collapse simulations comparing two 15 $M_{\&sun;}$ progenitor models with significant differences in initial Y_e (Woosley & Weaver 1995, Heger et al. 2000) exhibit no differences in Y_e at bounce, and, consequently, no difference in homologous core mass and shock formation radius. Fully dynamic simulations of core collapse, rebound, and shock propagation for 15 $M_{\&sun;}$ and 20 $M_{\&sun;}$ progenitor models of Nomoto & Hashimoto (1988) fail to produce explosions. In both cases, the shock stalls at 200 km, then recedes for several hundred milliseconds. The marked similarities observed in all these simulations highlight the need for both improved progesnitor models and the incorporation of improved microphysics in modern supernova codes. Spherically symmetric simulations are, for the immediate future, the only computationally feasible way to investigate the nature of the explosion mechanism while including the requisite level of detailed neutrino transport. They also provide one of the few opportunities to delineate the effects of various feedback mechanisms present in the problem. This research was supported by funds from the Joint Institute for Heavy Ion Research and a DOE PECASE award, and made use of the resources of the National Energy Research Scientific Computing Center, which is supported by the Office of Science of the U.S. DOE under Contract No. DE-AC03-76SF00098.

Spherically Symmetric Simulation with Boltzmann Neutrino Transport of Core Collapse and Postbounce Evolution of a 15 M_{solar} Star
Rampp, Markus; Janka, H.-Thomas
Astrophysical Journal, 539, 2000/8/1, L33-L36
Abstract - We present a spherically symmetric, Newtonian core collapse simulation of a 15 M_{solar} star with a 1.28 M_{solar} iron core. The time-, energy-, and angle-dependent transport of electron neutrinos (ν$_e$) and antineutrinos (ν&d1;$_e$) was treated with a new code that iteratively solves the Boltzmann equation and the equations for neutrino number, energy, and momentum to order O(v/c) in the velocity v of the stellar medium. The supernova shock expands to a maximum radius of 350 km instead of only ~240 km as in a comparable calculation with multigroup flux-limited diffusion (MGFLD) by Bruenn, Mezzacappa, & Dineva. This may be explained by stronger neutrino heating due to the more accurate transport in our model. Nevertheless, after 180 ms of expansion the shock finally recedes to a radius around 250 km (compared to ~170 km in the MGFLD run). The effect of an accurate neutrino transport is helpful but not large enough to cause an explosion of the considered 15 M_{solar} star. Therefore, postshock convection and/or an enhancement of the core neutrino luminosity by convection or reduced neutrino opacities in the neutron star seem necessary for neutrino-driven explosions of such stars. We find an electron fraction Y_e>0.5 in the neutrino-heated matter, which suggests that the overproduction problem of neutron-rich nuclei with mass numbers A~90 in exploding models may be absent when a Boltzmann solver is used for the ν$_e$ and ν&d1;$_e$ transport.

Spin-Flavour Conversions of Neutrinos in
Collapsing Stars
Husain, A.
New Worlds in Astroparticle Physics II,
1999/1/1, 244
Abstract - Not Available

Standard Physics Solution to the Solar Neutrino
Problem?
Dar, A.
New Worlds in Astroparticle Physics,
1998/1/1, 167
Abstract - Not Available

Stars as galactic neutrino sources
Brocato, E.; Castellani, V.; degl'Innocenti,
S.; Fiorentini, G.; Raimondo, G.
Astronomy and Astrophysics, 333,
1998/5/1, 910-917
Abstract - Theoretical expectations
concerning stars as neutrino sources are
presented according to detailed evaluations
of the stellar evolutionary histories for an
extended grid of stellar masses. Neutrino
fluxes and cumulative neutrino yields are
given for both 'thermo-nuclear' and
'cooling' neutrinos all over the nuclear life
of the stars and along the final cooling as
White Dwarfs. Predictions concerning the
galactic and the cosmic neutrino
background are presented and discussed.

Status and Perspectives of Neutrino Oscillation
Searches
Zuber, K.
Particles, Strings and Cosmology
(PASCOS 98), 1999/1/1, 280
Abstract - Not Available

Status and Perspectives of the Mainz Neutrino
Mass Experiment
Barth, H.; Borneschein, L.; Degen, B.;
Fleischmann, L.; Przyrembel, M.; Backe,
H.; Bleile, A.; Bonn, J.; Goldmann, D.;
Gundlach, M.; Kettig, O.; Otten, E. W.;
Tietze, G.; Weinheimer, Ch.; Leiderer, P.;
Kazachenko, O.; Kovalik, A.
Particle and Nuclear Physics, 1998/1/1,
353
Abstract - Not Available

Status of Neutrino Oscillation Searches
Zuber, K.

COSMO-97, First International Workshop
on Particle Physics and the Early Universe,
1998/1/1, 64
Abstract - Not Available

Status of the Milano neutrino mass experiment
with thermal detectors
Nucciotti, A. et al.
Weak Interactions and Neutrinos,
2000/1/1, 170
Abstract - Not Available

Status of the Neutrino Telescope AMANDA:
Monopoles and WIMPS
Rhode, W.
Dark Matter in Astro- and Particle Physics,
2001/1/1, 699
Abstract - Not Available

Status of Trimaximal Neutrino Mixing
Scott, W. G.
Identification of Dark Matter, 2001/1/1,
526
Abstract - Not Available

Sterile Neutrinos and CMB
Sciama, D. W.
AIP Conf. Proc. 476: 3K cosmology,
1999/1/1, 74
Abstract - Not Available

Sterile Neutrinos in Big Bang Nucleosynthesis
Buras, R.
Proceedings of the Sixth SFB-375
Ringberg Workshop Astroteilchenphysik,
2000/2/1, 82
Abstract - Not Available

Sterile Neutrinos: Phenomenology and Theory
Mohapatra, R. N.
AIP Conf. Proc. 478: COSMO-98,
1999/1/1, 440
Abstract - Not Available

Studies of neutrino oscillations at accelerators
Caldwell, David O.
Current aspects of neutrino physics,
2001/1/1, 155
Abstract - Contents: 1. Introduction. 2.
Motivation for the experiments. 3.
Intermediate-baseline ν_μ
experiments at high-energy accelerators. 4.
Short-baseline ν_μ and ν‾

μ experiments at lower-energy accelerators. 5. Conclusions.

Studies of neutrino oscillations at reactors
Boehm, Felix
Current aspects of neutrino physics, 2001/1/1, 131
Abstract - Contents: 1. Introduction. 2. The reactor neutrino spectrum. 3. Oscillation experiments. 4. The ν‾$_e$-d experiment at Bugey. 5. Neutrino magnetic moment. 6. Conclusion.

Studies of the Sudbury Neutrino Observatory detector and sonoluminescence using a sonoluminescent source
McDonald, Douglas Steven Ph.D. Thesis, 1999/1/1, 39
Abstract - The Sudbury Neutrino Observatory (SNO) is the first heavy water Cerenkov solar neutrino detector. 1000 metric tonnes of heavy water is used as a neutrino target and detection medium. SNO is designed to measure the flux and energy spectrum of high energy solar electron neutrinos via charged current interactions of electron neutrinos with deuterons in the heavy water. SNO can also measure the total high energy solar neutrino flux of neutrinos of all flavors via neutral current interactions with deuterons. The physics of solar models and solar neutrinos is presented. The physics of SNO and the SNO detector are described in detail. Two sonoluminescence sources were developed for use in calibrations of the SNO detector. The sonoluminescence source outperformed the standard SNO optical source, a nitrogen laser with a diffuser ball, by 25% in measurements of photomultiplier tube timing accuracy. Two systematic effects with the SNO electronics were discovered. Electronic crosstalk between channels has been measured for charges greater than 5 photoelectrons. An efficient cut has been developed to minimize this systematic error with an upper limit on signal loss of 0.4% times the PMT occupancy. Electronics crosstalk will affect solar neutrino analyses as it falsely adds PMT hits to some fraction of the events. A small anticorrelation of photomultiplier charges for electronics channels in close proximity has been measured. It has been shown that this subtle effect does not affect the number of photons detected, only the photomultiplier charges. A SL source and the SNO detector were used to study three properties of sonoluminescence. An analysis of SL data with regard to back-to-back photon correlations is presented. The results are consistent with no back-to-back photon correlations with an upper limit on the strength of back-to-back photon correlations of 3.16% for unfiltered SL light, and 0.5% for filtered SL light (λ = 420 nm) at the 95% confidence level. (Abstract shortened by UMI.)

Studies Towards a Detector for Extragalactic SuperNova Neutrinos
Stodolsky, L.
Proceedings of the Sixth SFB-375 Ringberg Workshop Astroteilchenphysik, 2000/2/1, 12
Abstract - Not Available

Study of Salt Neutrino Detector
Chiba, M.; Kamijo, T.; Kawaki, M.; Husain, A.; Inuzuka, M.; Ikeda, M.; Yasuda, O.
AIP Conf. Proc. 579: Radio Detection of High Energy Particles, 2001/1/1, 204
Abstract - Not Available

Study of the Process e^{+e-}-->W^{+W-} in a Model with Four Majorana Neutrinos
Teves, W. J. C.; Zukanovich Funchal, R.
Particles and Fields, Eighth Mexican School, 1999/1/1, 422
Abstract - Not Available

Sudbury Neutrino Observatory energy calibration using gamma-ray sources
Dragowsky, Michael Raymond Ph.D. Thesis, 1999/1/1, 16
Abstract - The long-standing Solar Neutrino Problem describes the disagreement between the observed and predicted solar neutrino flux. An extension to the electroweak model of particle physics predicts that neutrino flavor may change as a neutrino propagates from its source, and may account for the Solar

Neutrino Problem. The Sudbury Neutrino Observatory will make two measurements of the ⁸B neutrino flux from the Sun. The first is independent of neutrino flavor, allowing for the first time a definitive measurement of the total ₈B neutrino flux. The second measurement will be sensitive only to the ₈B electron-neutrino flux. Comparison of these two measurements will allow a determination of whether the electron-neutrinos produced in the Sun change their flavor as they propagate to the Earth. Precise energy calibration is required to ensure accurate measurements. Two energy calibration sources will be described that rely on the coincident emission of a β particle and a γ-ray. The sources make use of ¹⁶IN and ²⁴Na β- decay. In addition to energy calibration, the ²⁴Na source provides an assessment of the rate of photodissociation of deuterium that is important in understanding backgrounds for both solar neutrino measurements.

Super-Kamiokande 0. 07 eV Neutrinos in Cosmology: Hot Dark Matter and the Highest Energy Cosmic Rays
Gelmini, G. B.
Sources and Detection of Dark Matter and Dark Energy in the Universe, 2001/1/1, 525
Abstract - Not Available

Supernova Bounds on Neutrino Properties: A Mini-Review
Nunokawa, H.
AIP Conf. Proc. 415: Beyond the Standard Model. From Theory to Experiment, 1998/1/1, 284
Abstract - Not Available

Supernova II Neutrino Bursts and Neutrino Massive Mixing
Cline, David B.
The Role of Neutrinos, Strings, Gravity, and Variable Cosmological Constant in Elementary Particle Physics, 2001/1/1, 91
Abstract - Not Available

Supernova neutrino detection in Borexino
Cadonati, L.; Calaprice, F. P.; Chen, M. C.
Astroparticle Physics, 16, 2002/2/1, 361-372
Abstract - We calculated the expected neutrino signal in Borexino from a typical Type II supernova at a distance of 10 kpc. A burst of around 110 events would appear in Borexino within a time interval of about 10 s. Most of these events would come from the reaction channel ν^-_e+p-->e$^+$+n, while about 30 events would be induced by the interaction of the supernova neutrino flux on ^{12}C in the liquid scintillator. Borexino can clearly distinguish between the neutral-current excitations ^{12}C(ν,ν')$^{12C^*}$ (15.11 MeV) and the charged-current reactions ^{12}C(ν$_{e}$,e$^-$)^{12}N and ^{12}C(ν^-_e,e$^+$)^{12}B, via their distinctive event signatures. The ratio of the charged-current to neutral-current neutrino event rates and their time profiles with respect to each other can provide a handle on supernova and non-standard neutrino physics (mass and flavor oscillations).

Supernova Neutrino Opacity from Nucleon-Nucleon Bremsstrahlung and Related Processes
Hannestad, Steen; Raffelt, Georg
Astrophysical Journal, 507, 1998/11/1, 339-352
Abstract - Elastic scattering on nucleons, nuN --> Nnu, is the dominant supernova (SN) opacity source for mu and tau neutrinos. The dominant energy- and number-changing processes were thought to be nue^- --> e^-nu and nunu&d1;<-->e^+e^- until Suzuki showed that the bremsstrahlung process nunu&d1;NN<-->NN was actually more important. We find that for energy exchange, the related "inelastic scattering process" nuNN<-->NNnu is even more effective by about a factor of 10. A simple estimate implies that the nu_mu and nu_tau spectra emitted during the Kelvin-Helmholtz cooling phase are much closer to that of nu&d1;_e than had been thought previously. To facilitate a numerical study of the spectra formation we derive a scattering kernel that governs both

bremsstrahlung and inelastic scattering and give an analytic approximation formula. We consider only neutron-neutron interactions; we use a one-pion exchange potential in Born approximation, nonrelativistic neutrons, and the long-wavelength limit, simplifications that appear justified for the surface layers of an SN core. We include the pion mass in the potential, and we allow for an arbitrary degree of neutron degeneracy. Our treatment does not include the neutron-proton process and does not include nucleon-nucleon correlations.

Supernova Neutrino Oscillation in the Presence of Random Magnetic Field
Sahu, S.
AIP Conf. Proc. 531: Particles and Fields, 2000/1/1, 355
Abstract - Not Available

Supernova neutrinos.
Burrows, A.; Young, T.
Physica Scripta Volume T, 85, 2000/1/1, 127-131
Abstract - Not Available

Supernovae, Neutrinos, and Amateur Astronomers
Robinson, Leif J.
Sky and Telescope, 98, 1999/8/1, 30
Abstract - Not Available

Tau neutrinos in the Auger Observatory: a new window to UHECR sources
Bertou, X.; Billoir, P.; Deligny, O.; Lachaud, C.; Letessier-Selvon, A.
Astroparticle Physics, 17, 2002/5/1, 183-193
Abstract - The cosmic ray spectrum has been shown to extend well beyond 10^{20}eV. With nearly 20 events observed in the last 40 years, it is now established that particles are accelerated or produced in the universe with energies near 10^{21}eV at the production site. In all production models neutrinos and photons are part of the cosmic ray flux. In acceleration models (bottom-up models), they are produced as secondaries of the possible interactions of the accelerated charged particle; in direct production models (top-down models) they are a dominant fraction of the decay chain. In addition, hadrons above the GZK threshold energy will also produce, along their path in the Universe, neutrinos and photons as secondaries of the pion photo-production processes. Therefore, photons and neutrinos are very distinctive signatures of the nature and distribution of the potential sources of ultrahigh energy cosmic rays. In the following we describe the tau neutrino detection and identification capabilities of the Auger Observatory. We show that in the range 3×10^{17}-3×10^{20}eV the Auger effective aperture reaches a few tenths of km^2sr, making the Observatory sensitive to fluxes as low as a few tau neutrinos per km^2sryear. In the hypothesis of νμ -->ντ oscillations with full mixing, this sensitivity allows to a probe of the GZK cutoff as well as providing model independent constraints on the mechanisms of production of ultrahigh energy cosmic rays.

Technique for Direct eV-Scale Measurements of the Mu and Tau Neutrino Masses Using Supernova Neutrinos
Beacom, J. F.; Boyd, R. N.; Mezzacappa, A.
Physical Review Letters, 85, 2000/10/1, 3568-3571
Abstract - Early black hole formation in a core-collapse supernova will abruptly truncate the neutrino fluxes. The sharp cutoff can be used to make model-independent time-of-flight neutrino mass tests. Assuming a neutrino luminosity of 10^{52} erg/s per flavor at cutoff and a distance of 10 kpc, Super-Kamiokande can detect an electron neutrino mass as small as 1.8 eV, and the proposed OMNIS detector can detect mu and tau neutrino masses as small as 6 eV. We present the first technique with direct sensitivity to eV-scale mu and tau neutrino masses.

TeV Neutrinos and GeV Photons from Shock Breakout in Supernovae
Waxman, Eli; Loeb, Abraham
Physical Review Letters, 87, 2001/8/1, 1101
Abstract - Not Available

Textures for Neutrino Mass Matrices in Gauge
Theories
Lola, S.; Vergados, J. D.
Particle and Nuclear Physics, 1998/1/1, 71
Abstract - Not Available

Textures of Neutrino Mass Matrix with Large
Flavor Mixing
Tanimoto, M.
New Era in Neutrino Physics , 1998/1/1,
195
Abstract - Not Available

The ^{51}Cr and ^{90}Sr sources in BOREXINO
as tools for neutrino magnetic moment
searches
Ianni, A.; Montanino, D.
Astroparticle Physics, 10, 1999/5/1, 331-
338
Abstract - We calculate the event rates
induced by a ^{51}Cr nu_e source and by a
^{90}Sr-^{90}Y nu_e source in BOREXINO
through elastic scattering on electrons,
assuming a nonzero neutrino magnetic
moment mu_nu. We consider a source
activity of about 2 MCi and estimate the
solar nu ("source-off") background for
various oscillation scenarios.

The AMANDA Neutrino Experiment and its
Expansion to 1 Kilometer Dimension
Haizen, F.
ASP Conf. Ser. 141: Astrophysics From
Antarctica, 1998/1/1, 368
Abstract - Not Available

The AMANDA Neutrino Telescope
Bergström, L.; The AMANDA
Collaboration
The Identification of Dark Matter ,
1999/1/1, 501
Abstract - Not Available

The AMANDA Neutrino Telescope
Halzen, F.
COSMO-97, First International Workshop
on Particle Physics and the Early Universe,
1998/1/1, 100
Abstract - Not Available

The AMANDA neutrino telescope
Halzen, F.
New Astronomy Review, 42, 1998/9/1,
289-299
Abstract - Electronic Article Available
from Elsevier Science.

The AMANDA Neutrino Telescope
Halzen, F.
Particle and Nuclear Physics, 1998/1/1,
377
Abstract - Not Available

The AMANDA Neutrino Telescope: Design,
Construction, and Performance
Lowder, D. M.
ASP Conf. Ser. 141: Astrophysics From
Antarctica, 1998/1/1, 220
Abstract - Not Available

The AMANDA Neutrino Telescope: Science
Prospects and Performance at First Light
Halzen, F.
AIP Conf. Proc. 423: Fundamental
Particles and Interactions, Frontiers in
Contemporary Physics, 1998/1/1, 154
Abstract - Not Available

The AMANDA Neutrino Telescope: Status and
Latest Results
Lowder, Douglas M.
American Astronomical Society Meeting,
192, 1998/5/1
Abstract - The AMANDA collaboration
has successfully deployed a 10-string array
(AMANDA-B10) of 302 photomultiplier
tubes at the South Pole at a depth of 1.5 to
1.9 km. The array has been used to
measure the optical quality of the ice,
reconstruct tracks from high-energy cosmic
ray muons, and measure the energy of
electromagnetic cascades. Simulations and
early analyses of data indicate that the
array has an effective area of $\sim$ 10,000
m(2) for throughgoing neutrino-induced
muons, with an angular resolution of
2.5(deg) and good rejection (> 10(5))
of down-going cosmic ray muons. I will
describe the design and construction of the
detector, the methods used to simulate its
performance and to reconstruct muon
tracks, and the results to date; I will
include a progress report on the ongoing
analysis of a full year of data (500 GB)
from the detector. Finally, I will discuss
future work; this will include a description

of the AMANDA-II array, already begun with three strings in 1997-98, which will have an effective area of up to ~ 10(5) m(2) and angular resolution of ~ 1(deg) .

The AMANDA South Pole Neutrino
    Telescope: First Light
    Halzen, F.
    American Astronomical Society Meeting, 195, 1999/12/1
    Abstract - We will discuss the performance of natural Antarctic ice between 1 and 2 kilometer depths as a particle detector. We will present a preliminary analysis of the first year of data from a neutrino telescope which uses large volumes of ultra-transparent South Pole ice as a low-noise particle detector sensing the Cherenkov light from neutrino-induced muons and electrons. This instrument is monitoring the sky for neutrinos from supernovae and gamma ray bursts. We are already performing a first search for neutrino emission from the most energetic cosmic processes involving pulsars, black holes, active galactic nuclei and the like. The detector also has unique capabilities in searching for neutrino mass and dark matter. We will argue however that a high energy neutrino telescope should ultimately have an effective volume of order 1 kilometer cube, and will present AMANDA's ongoing and future expansion.

The Antares Demonstrator. Towards a High
    Energy Undersea Neutrino Telescope
    Blondeau, F.; ANTARES Collaboration
    Particle and Nuclear Physics, 1998/1/1, 413
    Abstract - Not Available

The ANTARES Neutrino Telescope: Status and
    Prospects
    Cartwright, S. L.
    Identification of Dark Matter, 2001/1/1, 506
    Abstract - Not Available

The atmospheric neutrino anomaly: muon
    neutrino disappearance
    Learned, John G.
    Current aspects of neutrino physics,
2001/1/1, 89
    Abstract - Contents: 1. Introduction. 2. The SuperKamiokande revolution. 3. Implications - astrophysics and cosmology, the theoretical situation: Why so important? Future muon neutrino experiments.

The Baikal Deep Under Water Neutrino
    Experiment: Results, Status, Future
    Balkanov, V. A.; Belolaptikov, I. A.; Bezrukov, L. B.; Budnev, N. M.; Chensky, A. G.; Danilchenko, I. A.; Djilkibaev, Zh.-A.; Domogatsky, G. V.; Doroshenko, A. A.; Fialkovsky, S. V.; Ganponeko, O. N.; Garus, A. A.; Gress, T. I.; Klabukov, A. M.; Klimov, A. I.; Klimushin, S. I.; Koshechkin, A. P.; Kulepov, V. F.; Kuzmichev, L. A.; Lovzov, S. V.; Lubsadorzhiev, B. K.; Milenin, M. B.; Mirgazov, R. R.; Moroz, A. V.; Moseiko, N. I.; Nikiforov, S. A.; Osipova, E. A.; Panfilov, A. I.; Parfenov, Yu. V.
    Particle and Nuclear Physics, 1998/1/1, 391
    Abstract - Not Available

The BAIKAL Neutrino Project: Status Report
    Balkanov, V. A. et al.
    Dark Matter in Astro- and Particle Physics, 2001/1/1, 707
    Abstract - Not Available

The C/CO ratio in B335 and the decaying dark
    matter neutrino hypothesis
    Murphy, B. T.; Little, L. T.; Kelly, M. L.
    Monthly Notices of the Royal Astronomical Society, 294, 1998/3/1, 635
    Abstract - CO-18 J = 2-1, CO-17 J = 2-1 and forbidden C I 3P1-3P0 emission from the dense cold cloud B335 has been observed and modelled in order to determine the C/CO ratio. The observed ratio is compared with a prediction by Tarafdar, who assumes a mechanism in which the CO dissociation is caused by photons of energy of about 13.8 eV. These were postulated by Sciama to result from the decay of dark matter neutrinos.

The Case for a Neutrino-Degenerate Universe
    Mathews, G. J.; Orito, M.; Kajino, T.; Wang, Y.

American Astronomical Society Meeting,
199, 2001/12/1
Abstract - We reanalyze the cosmological
constraints on the existence of a net
universal lepton asymmetry and neutrino
degeneracy based upon the latest high
resolution CMB sky maps from
BOOMERANG, DASI, and MAXIMA-1.
We compute likelihood functions for flat
cosmological models with and without
degenerate neutrinos. We adopt priors on
the degeneracy parameters and baryon
content based upon light-element
constraints on primordial nucleosynthesis.
We also adopt the prior of h = 0.72 +/- 0.08
from the Hubble Key Project results. Our
neutrino-degenerate CMB models include
a correction for that change in neutrino
decoupling temperature with degeneracy.

The Case of the Missing Neutrinos: Results
from SuperKamiokande
Haines, T. J.
AAS/High Energy Astrophysics Division,
31, 1999/4/1
Abstract - Neutrinos are enigmatic
particles that were invented in the 1930's
and discovered in the 1950's. Some 40
years after their discovery, much mystery
surrounds the properties of neutrinos.
Neutrinos come in three types, or "flavors,"
that describe the properties of their
interactions. It has long been speculated
that neutrinos may change flavor, so-called
neutrinos oscillations, as they propagate
through space. Data from the
SuperKamiokande experiment, located
under the Japanese Alps, has for the first
time produced evidence of neutrino
oscillations that is generally accepted. A
brief background about neutrino
oscillations will be given, followed by
many details about the data from
SuperKamiokande. A short discussion of
possible implications of neutrino
oscillations will close the talk.

The CERN Neutrino Oscillation Experiments
Burnner, J.
Particle and Nuclear Physics, 1998/1/1,
229
Abstract - Not Available

The CERN-LNGS neutrino programme
Pietropaolo, F.
Weak Interactions and Neutrinos,
2000/1/1, 416
Abstract - Not Available

The CUORE project: a large observatory for
neutrinoless double beta decay and other
rare events
Giuliani, A.; Cuore Collaboration
Lepton and Baryon Number Violation in
Particle Physics, neral Astrophysics and
Cosmology, 1999/1/1, 302
Abstract - Not Available

The Design and Status of the Sudbury Neutrino
Observatory
Cowen, D. F.
Eighteenth Texas Symposium on
Relativistic Astrophysics, 1998/1/1, 691
Abstract - Not Available

The Development and Calibration of An(127)I
Solar Neutrino Detector
Distel, James Ross Ph.D.
Thesis, 1998/10/1, 16
Abstract - We have measured the cross
section for the conversion of ^{127}I to ^{127}Xe
by electron neutrinos from the decay of
stopped muons as 2.54 ± 0.80 (stat)
± 0.14 (sys) ± 0.20 (flux/
uncertainty)$\times 10^{-40}$ cm^2. The measurement
was performed at the LAMPF proton
beamstop between 1993-95. A tank
containing 1540 kg of ^{127}I in the form of
NaI solution was placed 8.53 meters from
the beamstop where it received a neutrino
flux of order 5×10^7/ ν$_e$cm;^{2}/sec. The
^{127}Xe atoms that resulted from the capture
process were extracted from the target
solution, placed in miniature proportional
counters and counted. Our measured cross
section is in good agreement with the
theoretical prediction of Engel, et al., (2.1[-
]3.1$\times 10^{-40}$/ cm^2) and constitutes the first
determination of the capture cross section
from stopped muon decay neutrinos on a
nucleus to be used for (radiochemical)
solar neutrino detection. The measurement
is part of a larger effort to completely
calibrate ^{127}I with neutrinos across the
entire solar neutrino energy range; which
in the future will include a determination

of the capture cross section from ^{7}Be neutrinos using an intense ^{37}Ar source; and a determination of the differential cross section dσ&dE; on ^{127}I from stopped muon decay neutrinos in an electronic NaI detector. In addition to our calibration efforts, we have begun construction of a prototype iodine solar neutrino detector in the Homestake mine. The detector uses 100 tons of ^{127}I in the form of NaI solution. It is expected to have a signal comparable to that of the 615 ton chlorine detector and an order of magnitude smaller background. We have demonstrated that the detector has a rapid (~1 hour) extraction capability and improved event detection signature.

The east-west effect for atmospheric neutrinos
Lipari, P.
Astroparticle Physics, 14, 2000/11/1, 171-188
Abstract - The atmospheric neutrino fluxes observed by underground detectors are not symmetric under rotations around the vertical axis. The flux is largest (smallest) for the /ν's traveling toward east (west). This asymmetry is the result of the effects of the geomagnetic field on the primary cosmic rays and on their showers. The size of the asymmetry is to a good approximation independent of the existence of neutrino oscillations and can, in principle, be predicted without any ambiguity due to the existence of unknown new physics. We show that there is an interesting hint of discrepancy between the experimental measurement of the east-west asymmetry by Super-Kamiokande and the existing theoretical predictions. We argue that the discrepancy is a real effect caused by the neglect, in the existing calculations, of the bending of the charged secondary particles (particularly muons) in the cosmic ray showers. The inclusion of this effect in the calculation gives results in quantitative agreement with the data. We comment on the implications of this result for the study of neutrino oscillations.

The electron scattering reaction in the Sudbury Neutrino Observatory
Jillings, Christopher James Ph.D.
Thesis, 1999/1/1, 7
Abstract - The Sudbury Neutrino Observatory (SNO) is a 1000-tonne heavy-water Cerenkov detector designed to measure the flux and energy spectrum of electron-type solar neutrinos and the total flux of all flavours of solar neutrinos as well as neutrinos from other astrophysical sources. The angular resolution (the ability to measure the direction of recoil electrons from neutrino interactions with the heavy water) of the SNO detector has been studied in detail using Monte Carlo simulations. The factors affecting angular resolution, including the passage of electrons through water, the optical quality of detector components, and event reconstruction, are considered. The systematic uncertainty in angular resolution contributed by these factors is calculated. An analytic function which accurately describes the angular response of the SNO detector is presented. A method using a 6.13 MeV gamma-ray source to verify that the Monte Carlo calculations of angular resolution are correct is presented. The reconstructed direction for those events which reconstruct far from the source is compared to a measure of the initial direction of the gamma ray. The initial direction is taken to be the vector from the known location of the source to the reconstructed location of the event. Techniques to reduce the effect of misreconstruction of position are discussed. The impact of the angular resolution on the extraction of the electron-scattering (ES) signal is presented. Three simple analyses which use only directional information to extract the ES signal are presented. The statistical and systematic uncertainties are discussed in detail. The impact of the calibration on the ES extraction is discussed. Early detector data and calibrations are discussed with particular emphasis on use of the ^{16}N source before water was added to the detector and on gamma rays striking the water surface at partial fill. The behavior of the algorithms for reconstruction of direction during partial fill is presented.

The elusive neutrino - a subatomic detective
  story / Scientific American Library through
  W. H. Freeman, 1997 (Book Review)
  The Observatory, 118, 1998/1/1, 44
  Abstract - Not Available

The Evolution of Neutrino Astronomy
  Bahcall, John N.; Davis, Raymond, Jr.
  Publications of the Astronomical Society
  of the Pacific, 112, 2000/4/1, 429-433
  Abstract - This Essay is one of a series of
  invited contributions which will appear in
  the PASP throughout the year 2000 to
  mark the upcoming millennium. (Eds.)

The Extent and Cause of the Pre-White Dwarf
  Instability Strip, with Applications to
  Neutrino Astrophysics
  O'Brien, M. S.
  American Astronomical Society Meeting,
  193, 1998/12/1
  Abstract - The initial stages of white dwarf
  evolution are characterized by high
  luminosity, high effective temperature, and
  increasingly high surface gravity, making it
  difficult to constrain stellar properties
  through traditional spectroscopic
  observations. We are aided, however, by
  the fact that many pre-white dwarfs
  (PWDs) are multiperiodic g-mode
  pulsators. These stars fall into two classes,
  the variable planetary nebula nuclei
  (PNNV) and the "naked" GW Vir stars.
  Pulsations in PWDs provide a unique
  opportunity to probe their interiors, which
  are otherwise inaccesible to direct
  observation. Until now, however, the
  nature of the pulsation mechanism, the
  precise boundaries of the instability strip,
  and the mass distribution of the PWDs
  were complete mysteries. These problems
  must be addressed before we can apply
  knowledge of pulsating PWDs to improve
  understanding of white dwarf formation.
  This thesis lays the groundwork for future
  theoretical investigations of these stars. We
  first use Whole Earth Telescope
  observations to determine the mass and
  luminosity of the majority of the GW Vir
  pulsators. We find that pulsators of low
  mass have higher luminosity, suggesting
  the range of instability is highly mass-
  dependent. The observed trend of
  decreasing periods with decreasing
  luminosity matches a decrease in the
  maximum theoretical g-mode period
  accross the instability strip. We then show
  that the red edge can be caused by the
  lengthening of the driving timescale
  beyond the maximum sustainable period.
  This result is general for ionization-based
  driving mechanisms and explains the mass-
  dependence of the red edge. The form of
  the mass-dependence provides a vital
  starting point for future theoretical
  investigations of the driving mechanism.
  The blue edge probably remains
  undetected because of selection effects
  arising out of rapid evolution. Finally, we
  show that the observed rate of period
  change in cool GW Vir pulsators will
  constrain neutrino emission in their cores,
  and we identify appropriate targets for
  future observation.

The Extreme Energy Cosmic Rays and Cosmic
  Neutrinos as Probes for the Distant
  Universe. Astrophysics Involved and
  Experimental Approach.
  Scarsi, L.
  Current Topics in Astrofundamental
  Physics: the Cosmic Microwave
  Background, 2001/1/1, 499
  Abstract - Not Available

The Fermilab Neutrino Oscillation Facility
  Morfin, J. G.
  Particle and Nuclear Physics, 1998/1/1,
  239
  Abstract - Not Available

The fluxes of sub-cutoff particles detected by
  AMS, the cosmic ray albedo and
  atmospheric neutrinos
  Lipari, Paolo
  Astroparticle Physics, 16, 2002/1/1, 295-
  323
  Abstract - New measurements of the
  cosmic ray (c.r.) fluxes (p, e^+/- and helium)
  performed by the alpha magnetic
  spectrometer (AMS) during a 10 days
  flight of space shuttle have revealed the
  existence of significant fluxes of particles
  below the geomagnetic cutoff. These
  fluxes exhibit a number of remarkable
  properties, such as a ^3He ^4He ratio of order

/~10, an e^{+e^-} ratio of order /~4 and production from well-defined regions of the Earth that are distinct for positively and negatively charged particles. In this work we show that the natural hypothesis, that these sub-cutoff particles are generated as secondary products of primary c.r. interactions in the atmosphere can reproduce all the observed properties. We also discuss the implications of the sub-cutoff fluxes for the estimate of the atmospheric neutrino fluxes, and find that they represent a negligibly small correction. On the other hand the AMS results give important confirmations about the assumption of isotropy for the interplanetary c.r. fluxes also on large angular scales, and on the validity of the geomagnetic effects that are important elements for the prediction of the atmospheric neutrino fluxes.

The Galaxy Distribution and the Hubble Law in a Neutrino Dominated Universe
Sigismondi, C.; Filippi, S.; Ruffini, R.; Sanchez, L. A.
Memorie della Societa Astronomica Italiana, 69, 1998/1/1, 311
Abstract - Not Available

The geometry of atmospheric neutrino production
Lipari, P.
Astroparticle Physics, 14, 2000/11/1, 153-170
Abstract - The zenith angle distributions of atmospheric neutrinos are determined by the possible presence of neutrino oscillations and the combination of three most important contributions: (1) geomagnetic effects on the primary cosmic rays that suppress the primary flux in the Earth's magnetic equatorial region, (2) the zenith angle dependence of the neutrino yields due to the fact that inclined showers produce more neutrinos, and (3) geometrical effects due to the spherical shell geometry of the neutrino production volume. The last effect has been recognized only recently and results in an important enhancement of the flux of sub-GeV neutrinos for horizontal directions. In this work, we discuss the geometrical

effect and its relevance in the interpretation of the atmospheric neutrino data.

The Highest Energy Cosmics Rays, Photons, and Neutrinos
Zas, E.
AIP Conf. Proc. 445: Particles and Fields, Sixth Mexican Workshop, 1998/1/1, 185
Abstract - Not Available

The history of solar neutrino problem
Paterno, L.
Long and Short Term Variability in Sun's History and Global Change, collected and edited by Wilfried Schröder.
Interdivisional Commission on History of the IAGA, European Section. Newsletters of the Interd. Comm. History, No. 39, 2000, p.163, 2000/1/1, 163
Abstract - Not Available

The Interplay between Proto--Neutron Star Convection and Neutrino Transport in Core-Collapse Supernovae
Mezzacappa, A.; Calder, A. C.; Bruenn, S. W.; Blondin, J. M.; Guidry, M. W.; Strayer, M. R.; Umar, A. S.
Astrophysical Journal, 493, 1998/1/1, 848
Abstract - We couple two-dimensional hydrodynamics to realistic one-dimensional multigroup flux-limited diffusion neutrino transport to investigate proto-neutron star convection in core-collapse supernovae, and more specifically, the interplay between its development and neutrino transport. Our initial conditions, time-dependent boundary conditions, and neutrino distributions for computing neutrino heating, cooling, and deleptonization rates are obtained from one-dimensional simulations that implement multigroup flux-limited diffusion and one-dimensional hydrodynamics. The development and evolution of proto-neutron star convection are investigated for both 15 and 25 M&sun; models, representative of the two classes of stars with compact and extended iron cores, respectively. For both models, in the absence of neutrino transport, the angle-averaged radial and angular convection velocities in the initial Ledoux unstable region below the shock after

bounce achieve their peak values in ~20 ms, after which they decrease as the convection in this region dissipates. The dissipation occurs as the gradients are smoothed out by convection. This initial proto-neutron star convection episode seeds additional convectively unstable regions farther out beneath the shock.

The jet-disk symbiosis model for gamma ray bursts: cosmic ray and neutrino background contribution
Pugliese, G.; Falcke, H.; Wang, Y. P.; Biermann, P. L.
Astronomy and Astrophysics, 358, 2000/6/1, 409-416
Abstract - The relation between the cosmological evolution of the jet-disk symbiosis model for GRBs and the cosmic rays energy distribution is presented. We used two different Star Formation Rates (SFR) as a function of redshift and a Luminosity Function (LF) distribution to obtain the distribution in fluence of GRBs in our model and compare it with the data. We show a good agreement between the fluence distribution we obtain and the corrected data for the 4B BATSE catalogue. The results we obtain are generally valid for models that use jet physics to explain GRB properties. The fluence in the gamma ray band has been used to calculate the energy in cosmic rays both in our Galaxy and at extragalactic distances as a function of the redshift. This energy input has been compared with the Galactic and extragalactic spectrum of cosmic rays and neutrinos. Using our jet disk symbiosis model, we found that in both cases GRBs cannot give any significant contribution to cosmic rays. We also estimate the neutrino background, obtaining a very low predicted flux. We also show that the fit of our model with the corrected fluence distribution of GRBs gives strong constraints of the star formation rate as a function of the redshift.

The long baseline neutrino oscillation experiment from KEK to Super-Kamiokande - k2k
Suzuki, Y.; The K2k Collaboration
Weak Interactions and Neutrinos,
2000/1/1, 405
Abstract - Not Available

The nature of massive neutrinos
Kayser, Boris; Mohapatra, Rabindra N.
Current aspects of neutrino physics, 2001/1/1, 17
Abstract - Contents: 1. Introduction. 2. Dirac and Majorana masses for neutrinos. 3. Leptonic mixing and neutrino oscillation. 4. Experimental tests of the Majorana nature of the neutrino. 5. Neutrino decays.

The neutrino mass from tritium β decay
Otten, E. W.; Weinheimer, Ch.
Lepton and Baryon Number Violation in Particle Physics, neral Astrophysics and Cosmology, 1999/1/1, 309
Abstract - Not Available

The Neutrino Telescope ANTARES
Rostovtsev, A.
AIP Conf. Proc. 531: Particles and Fields, 2000/1/1, 351
Abstract - Not Available

The neutrino-induced neutron source in helium shell and r-process nucleosynthesis
Nadyozhin, D. K.; Panov, I. V.; Blinnikov, S. I.
Astronomy and Astrophysics, 335, 1998/7/1, 207-217
Abstract - The huge neutrino pulse that occurs during the collapse of a massive stellar core, is expected to contribute to the origination of a number of isotopes both of light chemical elements and heavy ones. In particular, evaporation of neutrons from helium nuclei excited by neutrino-nuclear inelastic collisions, may result in the r-process as it was first discussed by Epstein et al. (1988). Here we consider mainly the possibility to obtain the considerable amount of neutrons owing to the neutrino breakup of helium nuclei. It is shown that, in general, the heating of stellar matter due to the neutrino scattering off electrons and the heat released from the neutrino-helium breakup followed by the thermonuclear reactions should be taken into account.

The Neutrino-Oscillation Experiment at the
Chooz Nuclear Power Plant
Nicolò, D.
Particle and Nuclear Physics, 1998/1/1,
263
Abstract - Not Available

The Nuclear Physics of Solar and Supernova
Neutrino Detection
Haxton, W. C.
New Era in Neutrino Physics , 1998/1/1,
35
Abstract - Not Available

The Oscillation Length Resonance in the
Transitions of Solar and Atmospheric
Neutrinos Crossing the Earth Core
Petcov, S.
New Era in Neutrino Physics , 1998/1/1,
219
Abstract - Not Available

The Palo Verde Reactor Neutrino Experiment.
a Test for Long Baseline Neutrino
Oscillations
Boehm, F.; Hanson, J.; Henrikson, H.;
Michael, D.; Novikov, V. M.; Piepke, A.;
Vogel, P.; Yang, S.; Gratta, G.; Miller, L.;
Tracy, D.; Wang, Y. F.; Busenitz, J.;
Cornis, J.; Vital, A.; Wolf, J.; Dugger, M.;
Lawrence, D.; Ritchie, B.; Pittalwala, S.;
Wilfred, R.; Young, S.
Particle and Nuclear Physics, 1998/1/1,
253
Abstract - Not Available

The Palo Vere Neutrino Oscillation Experiment
Gratta, G.
Particles and the Universe, 1998/1/1, 366
Abstract - Not Available

The Pattern of Neutrino Masses and How to
Determine It
Barger, V.
AIP Conf. Proc. 478: COSMO-98,
1999/1/1, 433
Abstract - Not Available

The Predicted Signature of Neutrino Emission
in Observations of Pulsating Pre-White
Dwarf Stars
O'Brien, M. S.; Kawaler, S. D.
Astrophysical Journal, 539, 2000/8/1, 372-
378
Abstract - Pre-white dwarf (PWD)
evolution can be driven by energy losses
caused by neutrino emission in the core.
Unlike the solar neutrino flux, this is not
the by-product of nuclear fusion but is
instead the result of electron-scattering
processes in the hot, dense regions of the
PWD core. We show that the observed rate
of period change in cool PWD pulsators
will constrain neutrino emission in their
cores, and we identify appropriate targets
for future observation. Such a
measurement will tell us whether the
theories of lepton interactions correctly
describe the production rates and therefore
neutrino cooling of PWD evolution. This
would represent the first test of standard
lepton theory in dense plasma.

The Proposed ORLaND Neutrino Facility
Avignone, F. T., III; Efremenko, Yu. V.
Identification of Dark Matter, 2001/1/1,
578
Abstract - Not Available

The rate of cosmic ray showers at large zenith
angles: a step towards the detection of
ultra-high energy neutrinos by the Pierre
Auger Observatory
Ave, M.; Vázquez, R. A.; Zas, E.; Hinton,
J. A.; Watson, A. A.
Astroparticle Physics, 14, 2000/9/1, 109-
120
Abstract - It is anticipated that the Pierre
Auger Observatory can be used to detect
cosmic neutrinos of />10^{19}> eV that
arrive at very large zenith angles.
However, showers created by neutrino
interactions close to the detector must be
picked out against a background of similar
events initiated by cosmic ray nuclei. As a
step towards understanding this
background, we have made the first
detailed analysis of air showers recorded at
Haverah Park (an array which used similar
detectors to those planned for the Auger
Observatory) with zenith angles above
/60°. We find that the differential shower
rate from /60° to /80° can be predicted
accurately when we adopt the known
primary energy spectrum above 10^{17} eV
and assume the Quark Gluon String model

(QGSJET) and proton primaries. Details of
the calculation are given.

The Role of Neutrinos, Strings, Gravity, and
    Variable Cosmological Constant in
    Elementary Particle Physics
    Kursunoglu, Behram N.; Mintz, Stephan
    L.; Perlmutter, Arnold
    The Role of Neutrinos, Strings, Gravity,
    and Variable Cosmological Constant in
    Elementary Particle Physics, 2001/1/1
    Abstract - Not Available

The r-Process in Neutrino-driven Winds from
    Nascent, "Compact" Neutron Stars of Core-
    Collapse Supernovae ; Wanajo, Shinya;
    Kajino, Toshitaka; Mathews, Grant J.;
    Otsuki, Kaori
    Astrophysical Journal, 554, 2001/6/1, 578-
    586
    Abstract - We present calculations of r-
    process nucleosynthesis in neutrino-driven
    winds from the nascent neutron stars of
    core-collapse supernovae. A full dynamical
    reaction network for both the α-rich
    freezeout and the subsequent r-process is
    employed. The physical properties of the
    neutrino-heated ejecta are deduced from a
    general relativistic model in which
    spherical symmetry and steady flow are
    assumed. Our results suggest that proto-
    neutron stars with a large compaction ratio
    provide the most robust physical
    conditions for the r-process. Our results
    have confirmed that the neutrino-driven
    wind scenario is still a promising site in
    which to form the solar r-process
    abundances. However, our best results
    seem to imply both a rather soft neutron-
    star equation of state and a massive proto-
    neutron star that is difficult to achieve with
    standard core-collapse models. We propose
    that the most favorable conditions perhaps
    require that a massive supernova
    progenitor forms a massive proto-neutron
    star by accretion after a failed initial
    neutrino burst.

The r-Process Nucleosynthesis in Neutrino-
    /Magnetocentrifugally-Driven Winds
    Nagataki, Shigehiro; Kohri, Kazunori
    Publications of the Astronomical Society
    of Japan, 53, 2001/6/1, 547-553

Abstract - We consider the effects of the
rotation and magnetic fields on the rapid
neutron capture nucleosynthesis (r-process
nucleosynthesis) in neutrino-driven winds.
We examine the features of steady and
subsonic wind solutions which extend the
model of Weber and Davis (1967, Astron.
Jahresber. 67.6776), which is a
representative solar-wind model. Then, we
find that the entropy per baryon becomes
lower and the dynamical timescale
becomes longer as the angular velocity
becomes higher. The results are
inappropriate for the production of r-
process nuclei. Therefore, we conclude that
it is difficult to realize a successful r-
process nucleosynthesis in this framework.

The SIREN Solar Neutrino Experiment
    Liubarsky, I.; Bewick, A.; Sumner, T. J.;
    Marshall, R.; Blair, I. M.; Edgington, J. A.;
    Smith, N. J. T.; Smith, P. F.; Cartwright, S.
    L.; Lightfoot, P. K.; Kudryavtsev, V. A.;
    McMillan, J. E.; Spooner, N. J. C.
    Identification of Dark Matter, 2001/1/1,
    618

The solar neutrino problem after three hundred
    days of data at SuperKamiokande
    Fogli, G. L.; Lisi, E.; Montanino, D.
    Astroparticle Physics, 9, 1998/8/1, 119-130
    Abstract - We present an updated analysis
    of the solar neutrino problem in terms of
    both Mikheyev-Smirnov-Wolfenstein
    (MSW) and vacuum neutrino oscillations,
    with the inclusion of the preliminary data
    collected by the SuperKamiokande
    experiment during 306.3 days of operation.
    In particular, the observed energy spectrum
    of the recoil electrons from ^8B neutrino
    scattering is discussed in detail and is used
    to constrain the mass-mixing parameter
    space. It is shown that (1) the small mixing
    MSW solution is preferred over the large
    mixing one; (2) the vacuum oscillation
    solutions are strongly constrained by the
    energy spectrum measurement; and (3) the
    detection of a possible semiannual
    modulation of the ^8B nu flux due to
    vacuum oscillations should require at least
    one more year of operation of
    SuperKamiokande.

The Solar Neutrino Problem: Mixing of
Neutrinos and Mixing in the Sun
Haxton, W. C.
Particle and Nuclear Physics, 1998/1/1,
101
Abstract - Not Available

The Solar Neutrino Puzzle
Turck-Chieze, S.
Abstracts of the 19th Texas Symposium on
Relativistic Astrophysics and Cosmology,
held in Paris, France, Dec. 14-18, 1998.
Eds.: J. Paul, T. Montmerle, and E.
Aubourg (CEA Saclay)., 1998/12/1, 375
Abstract - After a rapid review of the
different activities which have contributed
to the understanding of solar neutrino
detections, I shall present very precise
helioseismic results which constrain the
solar structure and begin to reveal the solar
core. Sound speed, density and rotation
profiles are new interesting tools to explore
the central region in order to go beyond the
standard solar model. Therefore
helioseismology allows a better estimate of
the emitted neutrino fluxes and of their
related accuracy. Then I shall mention the
open questions and propose new directions
of improvements.

The solar neutrino puzzle: the way ahead
Turck-Chieze, S.
New Astronomy, 4, 1999/8/1, 325-332
Abstract - After a rapid review of the
different activities which have contributed
to the understanding of solar neutrino
detections, we examine the present
situation, the open questions and propose
new directions of improvements.

The standard solar model and the neutrino
problem: present status and future
perspectives
Paternò, Lucio
Memorie della Societa Astronomica
Italiana, 72, 2001/1/1, 483
Abstract - The inversion of thousands of 5-
minute acoustic modes, identified with
great accuracy, has clearly demonstrated
the correctness of the standard solar model.
The use of the most recent equation of state
and opacity tables together with the new
astrophysical factors for nuclear reactions

and the consideration of gravitational
settling of the elements heavier than
hydrogen during solar evolution has
permitted the construction of models which
are consistent with helioseismology within
some fraction of percent. This result
combined with the results of the four solar
neutrino experiments presently in operation
leads to the unavoidable conclusion that
the neutrino problem is not astrophysical in
origin. The solution should be searched in
new neutrino properties beyond those
considered in the minimal standard model
of electroweak interactions. Many
solutions with massive neutrinos have been
proposed, based on the change of neutrino
flavour during the Sun-Earth travel or
inside the Sun, which appear all promising
for solving the problem. The challenges of
the new millennium, in the first years of
which new experiments will produce new
results, are to search for the so called
"smoking-guns" or crucial tests for finding
the most consistent solution so
enlightening us about the neutrino
properties. Also laboratory experiments are
planned, which send a neutrino beam to a
detector placed at large distances crossing
the Earth's crust at depths of even 10 km,
to test directly the interaction of neutrino
with matter.

The Status of Neutrino Mass
Caldwell, D. O.
COSMO-97, First International Workshop
on Particle Physics and the Early Universe,
1998/1/1, 49
Abstract - Not Available

The status of neutrino physics
Valle, J. W. F.
Lepton and Baryon Number Violation in
Particle Physics, neral Astrophysics and
Cosmology, 1999/1/1, 111
Abstract - Not Available

The Sudbury Neutrino Observatory
McDonald, A. B.; SNO Collaboration
American Astronomical Society Meeting,
195, 1999/12/1
Abstract - The Sudbury Neutrino
Observatory (SNO) is a 1,000 tonne heavy
water Cerenkov detector situated 2,000

meters underground in INCO's Creighton mine near Sudbury, Ontario, Canada. The project is a Canadian, US and UK collaboration. Through the use of heavy water SNO will be able to detect a number of neutrino reactions, including one sensitive specifically to solar electron neutrinos and another to all active neutrino types. With these two reactions the detector will be able to search for neutrino flavor change without the requirement of electron neutrino flux normalization by solar model calculations. It will have a relatively high counting rate, on the order of 10 per day for solar neutrinos, and will also provide unusual sensitivity for measurements of other solar neutrino properties, atmospheric neutrinos and suprenova neutrinos. For supernova neutrinos, SNO will have high sensitivity for muon and tau neutrinos and anti-neutrinos as well as specific sensitivity for electron neutrinos and anti-neutrinos. It will have excellent timing and moderate directional sensitivity.

The sun as a high energy neutrino source
Hettlage, C.; Mannheim, K.; Learned, J. G.
Astroparticle Physics, 13, 2000/3/1, 45-50
Abstract - Cosmic ray interactions in the solar atmosphere yield a flux of electron and muon neutrinos with energies greater than 10 GeV. We discuss the influence of neutrino oscillations on the event rates in water-based Cerenkov detectors due to this neutrino flux and comment on the possibility of detecting the sun as a high energy neutrino source.

The Sun, a laboratory for neutrino- and astrophysics
Schlattl, H. Ph.D.
Thesis, 1999/1/1, 16
Abstract - Not Available

The Two Gravitating Massive Neutrino Pairs
Kursunoglu, Behram N.
Confluence , of Cosmology, Massive Neutrinos, Elementary Particles, and Gravitation, 1999/1/1, 5
Abstract - Not Available

The Weight of Neutrinos and Related Questions
Vissani, F.
Dark Matter in Astro- and Particle Physics, 2001/1/1, 435
Abstract - Not Available

Theoretical Possibilities and Observational Constraints for Radiatively Decaying Neutrinos with Mass Near 30 EV
Roos, M.
New Worlds in Astroparticle Physics, 1998/1/1, 226
Abstract - Not Available

Theory of neutrino masses and mixings
Mohapatra, Rabindra N.
Current aspects of neutrino physics, 2001/1/1, 217
Abstract - Contents: 1. Introduction. 2. Experimental indications of neutrino masses. 3. Patterns and textures for neutrinos. 4. Why neutrino mass necessarily means physics beyond the standard model. 5. Scenarios for small neutrino mass without right-handed neutrinos. 6. The seesaw mechanism and left-right symmetric unification models for small neutrino masses. 7. Naturalness of degenerate neutrinos. 8. Theoretical understanding of the sterile neutrino. 9. The E_6 model for the sterile neutrino. 10. The mirror universe model of the sterile neutrino. 11. Conclusions and outlook.

Thermodynamical instability of self-gravitating heavy neutrino matter
Bilic, N.; Viollier, R. D.
Dark matter in Astrophysics and Particle Physics, 1999/1/1, 684
Abstract - Not Available

Thermodynamics of the Neutrino Gas in a Plasma
Tsintsadze, L. N. et al.
New Worlds in Astroparticle Physics II, 1999/1/1, 236
Abstract - Not Available

Three Flavor Neutrino Oscillation Analysis of the Superkamiokande Atmospheric Neutrino Data
Yasuda, O.

New Era in Neutrino Physics , 1998/1/1,
165
Abstract - Not Available

Three-Neutrino Vacuum Oscillation Solutions
to the Solar and Atmospheric Anomalies
Whisnant, Kerry
Confluence of Cosmology, Massive
Neutrinos, Elementary Particles, and
Gravitation, 1999/1/1, 53
Abstract - Not Available

Time Variation of the Solar Neutrino Flux and
Correlations with Solar Phenomena
Stanev, T.
Eighteenth Texas Symposium on
Relativistic Astrophysics, 1998/1/1, 700
Abstract - Not Available

Top Five Reasons for Rejecting the Weyl-
Majorana Massless Neutrino Confusion
Theorem
Widom, A.
Particles, Strings and Cosmology
(PASCOS 98), 1999/1/1, 314
Abstract - Not Available

Towards the resolution of the solar neutrino
problem
Friedland, Alexander Ph.D.
Thesis, 2000/1/1, 9
Abstract - A number of experiments have
accumulated over the years a large amount
of solar neutrino data. The data indicate
that the observed solar neutrino flux is
significantly smaller than expected and,
furthermore, that the electron neutrino
survival probability is energy dependent.
This "solar neutrino problem" is best
solved by assuming that the electron
neutrino oscillates into another neutrino
species. Even though one can classify the
solar neutrino deficit as strong evidence for
neutrino oscillations, it is not yet
considered a definitive proof. Traditional
objections are that the evidence for solar
neutrino oscillations relies on a
combination of hard, different experiments,
and that the Standard Solar Model (SSM)
might not be accurate enough to precisely
predict the fluxes of different solar
neutrino components. Even though it
seems unlikely that modifications to the

SSM alone can explain the current solar
neutrino data, one still cannot completely
discount the possibility that a combination
of unknown systematic errors in some of
the experiments and certain modifications
to the SSM could conspire to yield the
observed data. To conclusively
demonstrate that there is indeed new
physics in solar neutrinos, new
experiments are aiming at detecting
"smoking gun" signatures of neutrino
oscillations, such as an anomalous seasonal
variation in the observed neutrino flux or a
day-night variation due to the regeneration
of electron neutrinos in the Earth. In this
dissertation we study the sensitivity reach
of two upcoming neutrino experiments,
Borexino and KamLAND, to both of these
effects.

Two-Neutrino Double Beta Decay: a Study of
Different Approximation Schemes
Simkovic, F.; Pantis, G.; Faessler, A.
Particle and Nuclear Physics, 1998/1/1,
285
Abstract - Not Available

Type-II supernovae and neutrino magnetic
moments
Nunokawa, H.; Tomàs, R.; Valle, J. W. F.
Astroparticle Physics, 11, 1999/7/1, 317-
325
Abstract - The present solar and
atmospheric neutrino data together with the
LSND results and the presence of hot dark
matter (HDM) suggest the existence of a
sterile neutrino at the eV scale. We have
reanalysed the effect of resonant sterile
neutrino conversions induced by neutrino
magnetic moments in a type-II supernova.
We analyse the implications of nu_e-nu_s
and nu_e-nu_s (nu_s denotes sterile
neutrino) conversions for the supernova
shock reheating, the detected nu_e signal
from SN1987A and the r-process
nucleosynthesis hypothesis.

Ultra High Energy Neutrinos by Tau
Airshowers
Fargion, D.
Dark Matter in Astro- and Particle Physics,
2001/1/1, 677

Ultra High Energy Neutrinos from Gamma Ray Bursts
Vietri, M.
The Active X-ray Sky: Results from BeppoSAX and RXTE, 1998/1/1, 694
Abstract - Not Available

Ultrahigh Energy Neutrinos as Probe for Weak-scale String Theories?
Kachelrieß, M.
20th Texas Symposium on relativistic astrophysics, 2001/1/1, 838
Abstract - Not Available

Ultrahigh energy neutrinos from gamma ray bursts.
Vietri, M.
Physical Review Letters, 80, 1998/1/1, 3690-3693
Abstract - Not Available

Ultra-High-Energy Neutrino Scattering onto Relic Light Neutrinos in the Galactic Halo as a Possible Source of the Highest Energy Extragalactic Cosmic Rays
Fargion, D.; Mele, B.; Salis, A.
Astrophysical Journal, 517, 1999/6/1, 725-733
Abstract - Diffuse relic neutrinos with light mass are transparent to ultra-high-energy (UHE) neutrinos at thousands of EeV, which are born through the photoproduction of pions by UHE protons on relic 2.73 K blackbody radiation (BBR), and originate in active galactic nuclei (AGNs) at cosmic distances. However, these UHE nu's may interact with others (mainly the heaviest: nu_mu_r, nu_tau_r, and respective antineutrinos) that are clustered into hot dark matter (HDM) galactic halos. UHE photons or protons, secondaries of nu-nu_r scattering, might be the final observed signatures of such high-energy chain reactions, and may be responsible for the highest energy extragalactic cosmic-ray (CR) events. Here we consider the conversion efficiency, ramifications, and energetics of these chain reactions for the 1991 October CR event at 320 EeV observed by the Fly's Eye detector in Utah. These quantities seem to be compatible with the distance, direction, and power (observed at MeV gamma energies) of the Seyfert galaxy MCG 8-11-11. The nu-nu_r interaction probability is favored by at least 3 orders of magnitude over a direct nu scattering onto Earth's atmosphere. Therefore, it may better explain the extragalactic origin of the puzzling 320 EeV event, while offering indirect evidence of a hot dark Galactic halo of light neutrinos (i.e., m_nu~tens of eV), probably of tau flavor.

Uncertainty of the solar neutrino energy spectrum
Liu, Q. Y.
Weak Interactions and Neutrinos, 2000/1/1, 222
Abstract - Not Available

Underwater-Ice Neutrino Telescopes
Racca, C.
New worlds in astroparticle physics, 2001/1/1, 93
Abstract - Not Available

Unitarity Constraints on Neutrino Mass and Mixings
Balantekin, A. B.
Particles, Strings and Cosmology, 2000/1/1, 423
Abstract - Not Available

Updated Parameters for the Decaying Neutrino Theory and E
Abstract - Various observational constraints are used to determine the domain of validity of the decaying neutrino theory for the ionisation of hydrogen in the interstellar medium of our Galaxy. The intensity of the decay line from neutrinos near the Earth, which is being searched for by E

Upper Bound on the Neutrino Magnetic Moment from Collisions Induced by Landau Damping in Supernovae
Ayala, A.
AIP Conf. Proc. 444: Particle Physics and Cosmology, First Tropical Workshop, 1998/1/1, 533
Abstract - Not Available

Vacuum energy and configuration energy for a massive neutrino propagating in a neutron

star
Woodahl, Brian Arvid Ph.D.
Thesis, 1999/1/1, 25
Abstract - The configuration energy of neutrons due to neutrino exchange at the one-loop level leads to an unphysically large energy in neutron stars when calculated under the constraints of the current Standard Model of massless neutrinos. However, the configuration energy can be made finite by endowing the three neutrino flavors with a minimum mass. The mass exponentially decreases the range of interaction among the neutrons in the star in the same manner as a Yukawa force is reduced. As the mass of the neutrino is increased the configuration energy of the neutron star is reduced to a physically acceptable value. A minimum neutrino mass can be determined from this calculation. We next contrast the configuration energy with the vacuum energy. The vacuum energy is obtained by computing the energy to all orders of a neutrino propagating in an external neutron field in the one-loop approximation. The vacuum energy does not include interactions among the neutrons themselves, it is the energy the vacuum itself acquires in the presence of the neutron field. The vacuum energy can be understood to be equivalent to the energy of the neutrino condensate that forms in the neutron star. We show that the vacuum energy or equivalently condensate energy increases as the neutrino mass increases. This occurs because massive neutrinos/antineutrinos increase the Fermi momentum which allows more particles to be trapped. Contrast this with the configuration energy which decreases as the neutrino mass is increased. We compute the vacuum energy to all orders and show its equivalence to the condensate energy. From this, it is demonstrated that both neutrinos and antineutrinos are trapped in a neutron medium. Also importantly, we show using a different formalism than Schwinger's logarithmic expansion how to compute the configuration energy of a neutron star. This new method of obtaining the configuration energy employs generalized derivative expansions. These higher-derivatives yield results identical to those obtained by Schwinger's formula. We point out that, previously unknown, the four-body configuration energy also has a dependence on r_c (the size of the neutron).

Vacuum oscillations and excess of high energy solar neutrino events observed in Superkamiokande
Berezinsky, V.
Astroparticle Physics, 12, 2000/1/1, 299-306
Abstract - Not Available

Vacuum oscillations and variations of solar neutrino rates in SuperKamiokande and Borexino
Faïd, B.; Fogli, G. L.; Lisi, E.; Montanino, D.
Astroparticle Physics, 10, 1999/1/1, 93-105
Abstract - The vacuum oscillation solution to the solar neutrino problem predicts characteristics variations of the observable neutrinos rates, as a result of the L/E_nu dependence of the nu_e survival probability (L and E_nu being the neutrino pathlength and energy, respectively). The E_nu-dependence can be studied through distortions of the recoil electron spectrum in the SuperKamiokande experiment. The L-dependence can be investigated through a Fourier analysis of the signal in the SuperKamiokande and Borexino experiments. We discuss in detail the interplay among such observable variations of the signal, and show how they can help to test and constrain the vacuum oscillation solution(s). The analysis includes the 374-day SuperKamiokande data.

Variability of the Solar Neutrino Flux
Sturrock, P. A.; Scargle, J. D.; Walther, G.; Weber, M. A.; Wheatland, M. S.
American Astronomical Society Meeting, 200, 2002/5/1
Abstract - Several tests of the available data provide evidence for variability of the solar neutrino flux. The variance of the Homestake measurements is larger than expected of a constant flux, and varies with heliographic latitude. The Homestake power spectrum contains a peak at 12.88 y⁻

[1] (period 28.4 days), corresponding to a sidereal rotation frequency of 440 nHz, close to that of the radiative zone. The power spectrum of GALLEX-GNO data contains the 12.88 y^{-1} peak and a stronger peak at 13.59 y^{-1} (period 26.9 days), corresponding to a sidereal rotation frequency of 462 nHz, that of the equatorial convection zone at normalized radius 0.85. Further evidence for time variation comes from the bimodality of the GALLEX-GNO and SAGE histograms. Joint spectrum analysis of the Homestake and GALLEX-GNO data yields evidence for the influence of r-mode oscillations [with l = 3, m = \{1,2,3\}] associated with the same sidereal rotation rate (13.88 y^{-1} or 440 nHz) found previously. The periods of these oscillations (158, 79, and 53 days, respectively) are close to those of known Rieger-type oscillations, and therefore point to the radiative zone as the source of these oscillations. A subset of these tests, selected to be independent, yield results that could arise by chance from a constant flux with probabilities ranging from 0.1 to 0.0001. If there are no relevant experimental systematic effects, and if the tests are valid and statistically independent, the combined estimates yield a probability of 10^{-15} that the results are compatible with a constant flux. A variable flux implies that neutrinos have a significant magnetic moment, and that neutrino measurements may be used to probe the Sun's internal magnetic field and internal dynamics. This work was supported by NASA grants NAS 8-37334 and NAG 5-9784, NSF grant AST-0097128, and the NASA Applied Information Systems Research Program.

Variations of the core luminosity and solar neutrino fluxes
Grandpierre, Attila
Proceedings of the 5th International Topical Workshop at LNGS on "Solar Neutrinos: Where Are the Oscillations", INFN, Laboratori Nazionali del Gran Sasso, March 12-14, 2001, edited by V. Berezinsky and F. Vissani, p. 487., 2001/1/1, 487
Abstract - The aim of the present work is to analyze the geological and astrophysical data as well as presenting theoretical considerations indicating the presence of dynamic processes present in the solar core. The dynamic solar model (DSM) is suggested to take into account the presence of cyclic variations in the temperature of the solar core.

Very High Energy Gamma-Rays and Neutrinos from AGN
Stecker, F. W.
American Astronomical Society Meeting, 192, 1998/5/1
Abstract - The AMANDA experiment and its proposed extension called ICECUBE are being built and planned to search for astrophysical sources of neutrinos and gamma-rays, particularly in energy range above 1 TeV. Gamma-rays in the 1 to 10 TeV range from three BL Lac objects have already been detected by air-Cerenkov telescopes. Mechanisms have been suggested by which neutrinos in this energy range and beyond may be generated either in the jets or cores of such blazars and other AGN. We will discuss such mechanisms and the source fluxes of gamma-rays and neutrinos which may be produced by AGN as well as the event rates which may be achieved in the Antarctic detectors.

VHE and UHE neutrinos from blazar flares
Rachen, J. P.; Meszaros, P.
Abstracts of the 19th Texas Symposium on Relativistic Astrophysics and Cosmology, held in Paris, France, Dec. 14-18, 1998. Eds.: J. Paul, T. Montmerle, and E. Aubourg (CEA Saclay)., 1998/12/1, 431
Abstract - We discuss limits on neutrino energy and fluxes from transient outbursts or flares in proton blazars, considering particle acceleration and loss processes of protons, but also losses of secondary particles in the photohadronic cascade leading to neutrino production. We show that neutrinos from blazar flares, and probably from non-transient emission as well, are essentially limited to energies below $10^{\{19\}}$ eV.

VHE and UHE neutrinos from GRB - energy and flux limits

Rachen, J. P.; Mucke, A.; Meszaros, P.
Abstracts of the 19th Texas Symposium on Relativistic Astrophysics and Cosmology, held in Paris, France, Dec. 14-18, 1998. Eds.: J. Paul, T. Montmerle, and E. Aubourg (CEA Saclay)., 1998/12/1, 79
Abstract - Discussing proton acceleration, photohadronic interaction physics and various energy loss processes of protons and secondary charged particles in the hadronic cascade leading to neutrino production in GRB, we set limits on possible contributions to the VHE and UHE neutrino spectrum. We show that neutrinos from internal or external shocks of GRB and their afterglows are unlikely to exceed the energy of $10^{\{19\}}$ eV with a measurable flux.

Weak Interactions and Neutrinos
Dominguez, C. A.; Viollier, R. D.
Weak Interactions and Neutrinos, 2000/1/1
Abstract - Not Available

Weighing neutrinos: weak lensing approach
Cooray, Asantha R.
Astronomy and Astrophysics, 348, 1999/8/1, 31-37
Abstract - We study the possibility for a measurement of neutrino mass using weak gravitational lensing. The presence of non-zero mass neutrinos leads to a suppression of power at small scales and reduces the expected weak lensing signal. The measurement of such a suppression in the weak lensing power spectrum allows a direct measurement of the neutrino mass, in contrast to various other experiments which only allow mass splittings between two neutrino species. Making reasonable assumptions on the accuracy of cosmological parameters, we suggest that a weak lensing survey of 100 sqr. degrees can be easily used to detect neutrinos down to a mass limit of ~ 3.5 eV at the 2sigma level. This limit is lower than current limits on neutrino mass, such as from the Lyalpha forest and SN1987A. An ultimate weak lensing survey of pi steradians down to a magnitude limit of 25 can be used to detect neutrinos down to a mass limit of 0.4 eV at the 2sigma level, provided that other cosmological parameters will be known to an accuracy expected from cosmic microwave background spectrum using the MAP satellite. With improved parameters estimated from the PLANCK satellite, the limit on neutrino mass from weak lensing can be further lowered by another factor of 3 to 4. For much smaller surveys (~ 10 sqr. degrees) that are likely to be first available in the near future with several wide-field cameras, the presence of neutrinos can be safely ignored when deriving conventional cosmological parameters such as the mass density of the Universe. However, armed with cosmological parameter estimates with other techniques, even such small area surveys allow a strong possibility to investigate the presence of non-zero mass neutrinos.

What can long-baseline neutrino oscillation experiments tell us about neutrino dark matter?
Minakata, Hisakazu; Shimada, Yoriaki
Astroparticle Physics, 8, 1998/2/1, 193-200
Abstract - We define the ratio R = (No. of observed electron events)/(No. of expected muon events - No. of observed muon events) as a measure for the relative importance of nu_mu-->nu_e and nu_mu-->nu_tau oscillations which will be measurable in coming long-baseline neutrino experiments. We then argue that if neutrino oscillation is observed with 0.02 < R < 0.87 it implies the rejection of the neutrino dark matter hypothesis in a wide range of the mixing parameters which is consistent with the atmospheric neutrino anomaly. Our argument is valid if neutrinos come with only three flavors without steriles and if they have a hierarchical mass pattern.

With Neutrino Masses Revealed, Proton Decay is the Missing Link
Pati, Jogesh C.
The Abdus Salam Memorial Meeting, 1999/1/1, 98
Abstract - Not Available

AUTHOR INDEX

A

Abazajian, Kevork Nazar Ph.D., 44
Adam, Shaffique, 83
Adams, J. A., 17
Adams, J., 17, 55
Ahrens, J., 81
Akhmedov, E. K., 54, 77
Aksenov, A. G., 16
Akulov, V. P., 70
Alessandrello, Angelo, 1
Alimonti, G., 79
Alvarez-Muñiz, J., 23, 74
Ambrosio, M., 43, 75
Amram, P., 10
Ando, S., 20
Andres, E., 39, 64
Añez, N. Y., 46
ANTARES Collaboration, 95
Antia, H. M., 12, 83
Antia, H., 12, 32, 83
Antolini, R., 43, 75
Anvar, S., 10
Apollonia, M., 27
Apostolovska, G. et al., 55
Arhipova, N. A., 5
Armbruster, B., 52
Arpesella, C., 79
Arras, P., 12, 75
Asano, Katsuaki, 53, 77
Askebjer, P., 39, 64
Aslanides, E., 10
Athar, H., 45
Atoyan, A., 34, 60
Aubert, J.-J., 10
Auriemma, G., 43, 75
Ave, M., 101
Avignone, F. T., III, 101

Ayala, A., 50, 106
Azoulay, R., 10

B

Back, H., 79
Backe, H., 90
Bahcall, J. N., 7, 34, 35, 42, 85, 86, 87, 98
Bahcall, J., 7, 35, 86, 87
Bai, X., 64, 81
Bakari, D., 43
Baker, R., 75
Balantekin, A. B., 106
Balata, M., 79
Baldini, A., 43, 75
Balkanov, V. A., 7, 76, 95
Barbarino, G. C., 43, 75
Barger, V., 28, 51, 68, 101
Barish, B. C., 43, 75
Barouch, G., 64, 81
Barshay, Saul, 51
Barth, H., 90
Barwick, S. W., 39, 64, 81
Basa, S., 10, 33
Basu, D., 86
Basu, S., 35, 85
Batkin, I. S., 13
Battistoni, G., 1, 43, 75
Baudis, L., 21
Baumgarte, Thomas W., 75
Bay, R. C., 64, 81
Bay, R., 39, 64, 81
Bazarko, Andrew O., 6
Beacom, J. F., 12, 38, 93
Beall, J. H., 60
Beau, T., 79
Beck, M., 4
Becka, T., 81
Becker, K.-H., 64, 81

Becker, M., 52
Bednarek, W., 26, 30, 60, 73
Beeman, Jeffrey W., 1
Belanger, G., 36
Bellini, G., 79
Bellotti, E., 85, 86
Bellotti, R., 43, 75
Belolaptikov, I. A., 7, 76, 95
Bemporad, C., 43, 75
Benen, A., 52
Benhammou, Y., 10
Bento, L., 53, 76
Benziger, J., 79
Berezhiani, Z., 26
Berezinsky, V., 18, 33, 87, 107
Berger, Thomas K., 83
Bergström, L., 34, 39, 64, 94
Bernabei, R., 72
Bernabéu, J., 19
Bernard, F., 10
Bernardini, P., 24, 43, 54, 75
Berthomieu, G., 85
Bertin, V., 10
Bertou, X., 93
Bertrand, D., 64, 81
Bethe, H. A., 13
Bewick, A., 102
Bezrukov, L. B., 7, 76, 95
Bharadwaj, Somnath, 20
Bierenbaum, D., 64
Biermann, P. L., 100
Bilenkey, S. M., 51
Bilenky, S. M., 28, 51
Bilic, N., 32, 104
Billault, M., 10
Biller, Steve, 36
Billoir, P., 93
Bilokon, H., 43, 75
Bingham, R., 13, 49, 69
Biron, A., 39, 64, 81
Bisi, V., 43, 75
Blair, I. M., 52, 102
Blanc, F., 10
Blanc, P.-E., 10
Bland, R. W., 10
Blasone, M., 32
Bleile, A., 90
Bleve, C., 24
Blinnikov, S. I., 100
Bloise, C., 43, 75
Blondeau, F., 10, 95
Blondin, J. M., 6, 41, 99
Bludman, S. A., 12

Bocean, Virgil, 6
Bodmann, B. A., 52
Boehm, F., 91, 101
Boettcher, M., 60
Boger, J., 22
Bonetti, S., 79
Bonn, J., 62, 90
Booth, J., 39, 64, 81
Booth, N. E., 52
Bordes, J., 70
Bordes, José, 70
Borexino Collaboration, 11, 79, 86
Borneschein, L., 90
Botner, O., 64, 81
Bottu, N., 10
Bouchta, A., 39, 64, 81
Boulesteix, J., 10
Bower, C., 43, 75
Bowyer, Stuart, 25
Boyce, M. M., 64, 81
Boyd, R. N., 5, 74, 93
Brabec, V., 40
Bregadze, V. I., 25
Brigatti, A., 79
Brigida, M., 43
Brocato, E., 90
Brofferio, Chiara, 1
Brooks, B., 10
Brudanin, V. B., 25
Bruenn, S. W., 1, 6, 22, 41, 57, 72, 99
Brun, A. S., 70, 85
Brunner, J., 10
Buckingham, Michael J., 83
Budnev, N. M., 7, 76, 95
Buras, R., 90
Burgess, C. P., 55
Burigana, C., 65
Burkart, R. F., 48
Burnner, J., 96
Burrows, A., 3, 5, 45, 59, 93
Busenitz, J., 101
Bussino, S., 43, 75

C

Caccianiga, B., 29, 58, 79
Cadonati, L., 79, 92
Cafagna, F., 43, 75
Calaprice, F., 79, 92
Calder, A. C., 6, 99
Caldwell, D. O., 5, 19, 46, 50, 59, 90, 103
Calicchio, M., 43, 75
Calzas, A., 10

Campana, D., 43, 75
Capelle, K. S., 65
Capozziello, S., 9
Carboni, M., 43, 75
Cardall, C. Y., 54, 65
Carius, S., 39, 64, 81
Carloganu, C., 10
Carlson, M., 39
Carmona, E., 10
Carr, J., 10
Carton, P.-H., 10
Cartwright, S., 10, 95, 102
Cases, R., 10
Casper, D. W., 10
Cassol, F., 10
Castellani, V., 85, 90
Castellano, M., 75
Cecchet, G., 79
Cecchini, S., 43, 75, 81
Cei, F., 43, 45, 75
Celio, P., 75
Chan, Hong-Mo, 70
Chechetkin, V. M., 83
Chen, A., 64, 81
Chen, M., 79, 92
Chen, Xuelei, 37
Chensky, A. G., 7, 76, 95
Chiarella, V., 43, 75
Chiba, M., 91
Chirkin, D., 64, 81
Chitre, S., 12, 32, 83
Choubey, S., 5, 37
Choudhary, B. C., 43
Christensen-Dalsgaard, J., 32, 85
Churkina, G. P., 83
Circella, M., 42
Cleveland, Bruce T., 40
Cline, D. B., 14, 74, 81, 92
Co', G., 24
Colgate, S., 74
Compere, C., 10
Conner, Z., 84
Conrad, J., 64, 81
Cooley, J., 64, 81
Cooper, S., 10
Cooray, Asantha R., 109
Cornis, J., 101
Corona, A., 75
Costa, C. G. S., 64, 81
Coustillier, G., 10
Coutu, S., 43, 75
Couvidat, S., 85
Cowen, D., 39, 64, 81, 96

Cremonesi, Oliviero, 1
Crocker, Roland M., 68
Croft, R. A. C., 18
Cronin, J. W., 65
Crooks, James L., 77
Cumming, J. B., 22
Cuore Collaboration, 96

D

Dai, Z. G., 42
Dailing, J., 64
Daily, Timothy, 40
Dalberg, E., 39, 64, 81
Dalhed, H. E., 29
Danilchenko, I. A., 7, 76, 95
Danilov, M., 20
Dar, A., 90
Davé, R., 18
Davis, Raymond, Jr., 40, 98
Dawson, J. M., 13, 49, 63, 69
de Benedictis, L., 75
de Botton, N., 10
De Cataldo, G., 43, 75
de Kerret, H., 79
De Marzo, C., 43, 75
De Mitri, I., 24, 43, 75
De Paolis, F., 9
De Vincenzi, M., 43, 75
DeBari, A., 79
Deck, P., 10
Dedenko, L. G., 74
Degen, B., 90
degl'Innocenti, S., 3, 85, 90
DeHaas, E., 79
Dekhissi, H., 43, 75
Deligny, O., 93
Derkaoui, J., 43, 67
Dermer, C. D., 34, 60
Desages, F. E., 10
Desiati, P., 64, 81
Destelle, J.-J., 10
Deutsch, M., 79
Devoe, R., 20
Dewulf, J.-P., 64, 81
DeYoung, T., 39, 63, 64, 81
Di Bari, P., 14
di Credico, A., 34, 43, 75
di Lella, L., 35
Dick, K., 19
Dienes, K. R., 69
Dienes, Keith R., 62
Dieperink, A. E. L., 58

Dietz, A., 21
Dimmock, J. O., 31
Dispau, G., 10
Distel, James R., 40, 96
Djilkibaev, Zh.-A., 76, 95
Doksus, P., 64, 81
Dokuchaev, V. I., 33
Dolgolenko, A., 20
Dolgov, A. D., 39, 46, 52, 53, 63
Dolgov, A. G., 61
D'Olivo, J. C., 50
Dominguez, C. A., 109
Domogatsky, G. V., 7, 76, 95
Domokos, G., 3, 64
Donghi, O., 79
Doroshenko, A. A., 7, 76, 95
Dragoun, O., 40
Dragounová, N., 40
Dragowsky, Michael Raymond Ph.D., 91
Drexlin, G., 52
Drogou, J. F., 10
Drouhin, F., 10
Dubeau, J., 3
Dugger, M., 101
Duley, W. W., 2, 19
Dunn, James O., 77
Duval, P.-Y., 10
Dvornikov, M., 68
Dzhilkibaev, Z.-A. M., 7
Dziembowski, W. A., 85

E

Eastman, Ron, 3
Eberhard, V., 52
Ebisuzaki, T., 23, 31
Edelstein, Jerry, 25
Edgecock, R., 3
Edgington, J. A., 52, 102
Edsjö, J., 64, 81
Efremenko, Yu. V., 3, 101
Egorov, A. M., 24
Eichner, C., 52
Eitel, K., 37, 52
Ejiri, H., 59, 71
Ekers, R. D., 4
Ekström, P., 39, 64, 81
Elisei, F., 79
Enriquez, O., 43
Erlandson, B., 39
Erlandsson, B., 64
Erriquez, O., 75
Etenko, A., 79

F

Faessler, A., 105
Faïd, B., 107
Falcke, H., 100
Faraggi, A. E., 72
Fargion, D., 20, 22, 29, 79, 83, 105, 106
Faridani, J., 70
Favuzzi, C., 43, 75
Fazely, Ali R., 72
Fedorova, A. V., 36
Feinstein, F., 10
Fenyves, E., 21, 48, 74
Fernholz, R., 79
Ferrari, N., 85
Feser, T., 64, 81
Festy, D., 10
Fetter, Jonathan Morton Ph.D., 78
Fialkovsky, S. V., 7, 76, 95
Filippi, S., 99
Filossofov, D. V., 25
Finckh, E., 52
Finelli, F., 65
Fiorentini, G., 5, 85, 87, 90
Fiorillo, G., 76
Fiorini, Ettore, 1
Fleischmann, L., 90
Fogli, G. I., 4, 53
Fogli, G. L., 4, 102, 107
Font, J. A., 88
Foot, R., 14, 36
Foote, Richard H., 6
Fopma, J., 10
Ford, R., 11, 79
Fornengo, N., 51, 55
Forti, C., 43, 75
Frampton, Paul H., 77
Freudiger, B., 79
Freund, M., 69
Frichter, George M., 42
Friedland, Alexander Ph.D., 105
Fritzsch, H., 42
Fuda, J.-L., 10
Fujita, T., 76
Fukasaku, K., 76
Fukuyama, Takeshi, 53, 77
Fulgione, W., 32
Fuller, G. M., 48, 51, 59, 64, 65, 74
Fusco, P., 43, 75
Fushimi, K., 71

G

Gabriel, A. H., 85
Gaisser, T. K., 71, 83
Gaitán, R., 11
Galbiati, C., 79
Gaponenko, O. N., 7, 76, 95
Garagiola, A., 79
García, E., 11
García, R. A., 70, 85
Garus, A. A., 76, 95
Garvey, G. T., 82
Garzelli, M. V., 2
Gatti, F., 79
Gaug, M., 64, 81
Gazzana, S., 79
Gelb, J. M., 8, 82
Gelmini, G., 61, 62, 92
Gemmeke, H., 52
Giacomelli, G., 43, 75, 81
Giammarchi, M., 11, 58, 79
Giannini, G., 20, 27, 43, 75
Giglietto, N., 43, 75
Giorgini, M., 43
Giugni, D., 79
Giuliani, A., 1, 96
Giunti, C., 2, 28, 51
Glück, M., 22
Gnedin, Nickolay Y., 19
Gnedin, O. Y., 19, 48
Goldmann, D., 90
Goldschmidt, A., 64, 81
Golubchikov, A., 79
Gómez, José F., 25
Gonzalez-Garcia, M. C., 7
González-Romero, L. M., 19
Goobar, A., 39, 64
Goret, P., 10
Goretti, A., 79
Gorham, P. W., 62, 66
Gosset, L., 10
Goswami, S., 5, 37, 83
Gough, D. O., 85, 87
Gournay, J.-F., 10
Grandpierre, Attila, 2, 108
Grassi, M., 43, 74, 75
Grasso, D., 53
Gratta, G., 20, 51, 83, 101
Gray, L., 39, 43, 64, 75
Gress, T. I., 76, 95
Grieb, C., 79
Grillo, A., 43, 75
Grimus, W., 27, 28, 51

Grossi, M., 29
Guarino, F., 43, 75
Guarnaccia, P., 75
Guetta, D., 49, 66
Guidry, M., 1, 6, 55, 57, 99
Gundlach, M., 90
Gundorin, N. A., 25
Gusev, A. V., 68
Gustavino, C., 43, 75
Guzzo, M. M., 20

H

Haase, H., 64
Habig, A., 43, 75, 84
Hadaway, J. B., 31
Haensel, P., 48
Hagmann, C., 4
Hagner, C., 79
Hagner, T., 79
Hahn, R. L., 22
Haines, T. J., 96
Haizen, F., 94
Haller, Eugene E., 1
Hallgren, A., 39, 64, 81
Halprin, A., 6
Halzen, F., 33, 34, 35, 39, 43, 48, 64, 81, 94, 95
Hampel, W., 79
Hankins, T. H., 4
Hannestad, Steen, 92
Hansen, S. H., 17, 39, 53, 63
Hanson, J., 101
Hanson, K., 3, 43, 64, 75, 81
Harding, E., 79
Hardtke, R., 39, 64, 81
Hargrove, C. K., 3
Hart, S., 39
Hartmann, D., 54
Hartmann, F., 79
Hasegan, D., 81
Haug, S., 52
Hawthorne, A., 75
Haxton, W., 53, 54, 84, 101, 103
Hayashi, K., 71
Hazama, R., 71
He, Y., 39, 64
Heeger, K. M., 3
Heinz, R., 43, 75
Hellwig, M., 64, 81
Hencheck, M., 74
Henning, P., 32
Henrikson, H., 101
Herin, J., 52

Hernández, J. J., 10
Hernández-Galeana, A., 11
Herrouin, G., 10
Hess, H., 79
Hettlage, C., 35, 104
Heukenkamp, H., 39, 64, 81
Heusser, G., 21, 79
Hill, G., 39, 64, 81
Hillman, L. W., 31
Hime, A., 74
Hinton, J. A., 101
Hirsch, M., 2, 84, 87
Hix, W. R., 22, 31, 41, 72
Hoffman, R., 54
Honda, M., 71
Hong, J. T., 75
Hong-Mo, C., 70
Hooper, D. W., 48
Hosogai, M., 7
Hößl, J., 52
Hoyle, Dixon, 6
Hu, W., 18
Hubaut, F., 10
Hubbard, J. R., 10
Hulth, P. O., 39, 64, 81, 82
Hundertmark, S., 39, 64, 81
Husain, A., 90, 91
Huss, D., 10

I

Ianni, A., 79, 94
Iarocci, E., 43, 75
Ikeda, M., 91
Imlay, R. I., 51
Ingrosso, G., 9
Inuzuka, M., 91
Inzani, P., 79
Iovane, G., 9
Iyer, Sharada Ramlingam Ph.D., 4
Ízek, J., 40

J

Jacobsen, J., 39, 64, 81
Jain, Pankaj, 42
Janka, H. T., 16, 26, 56, 59, 88
Janka, H. Th., 13, 16, 56, 89
Jannakos, T., 52
Jaquet, M., 10
Jelley, N., 10
Jillings, Christopher James Ph.D., 97
Jochum, J., 13

Jones, A., 39
Jünger, P., 52

K

Kachelrieß, M., 106
Kahniashvili, T., 5
Kajfasz, E., 10
Kajino, T., 31, 46, 62, 95, 102
Kajita, T., 10, 50, 84
Kamijo, T., 91
Kaminker, A. D., 47, 48, 58
Kaminski, W. A., 17
Kandhadai, V., 39, 64
Kaneyuki, K., 78
Kar, Kamales, 5, 83
Karle, A., 39, 64, 81
Karlik, Y. S., 74
Karolak, M., 10
Katsavounidis, E., 43, 75
Katsavounidis, I., 43
Kawaki, M., 91
Kawaler, S. D., 101
Kawasaki, M., 71
Kayser, B., 14
Kayser, Boris, 100
Kazachenko, O., 90
Kearns, E., 43, 75
Kelly, M. L., 95
Keranen, Petteri Ph.D., 9
Kettig, O., 90
Khlopov, M., 29
Kidner, S., 79
Kieda, D., 80
Kielczewska, D., 59
Kiko, J., 79
Kim, H., 43
Kim, J., 50, 64, 81
Kirilova, D. P., 71
Kirsten, T., 30, 79
Kishimoto, T., 71
Kiss, D., 7
Klabukov, A. M., 7, 76, 95
Klapdor-Kleingrothaus, H. V., 3, 21, 31, 58, 84, 87
Kleifges, M., 52
Klein, Joshua R., 74
Kleinfeller, J., 52
Klimov, A. I., 7, 76, 95
Klimushin, S. I., 7, 76, 95
Klochek, N., 38
Kluzniak, W., 52
Koci, B., 39, 64, 81

Kodama, K., 20
Kohri, K., 71, 102
Koike, M., 19
Kolb, S., 13, 84
Komori, M., 71
Konoplich, R., 29
Köpke, L., 64, 81
Korga, G., 79
Korpela, Eric J., 25
Korschinek, G., 79
Koshechkin, A. P., 7, 76, 95
Kosovichev, A. G., 85
Kouchner, A., 10
Kovalenko, S. G., 2
Kovalík, A., 40, 90
Kovesi-Domokos, S., 3, 64
Kowalski, M., 64, 81
Kretschmer, W., 52
Kretzer, S., 22
Kreyerhoff, Georg, 51
Królikowski, W., 8
Kryn, D., 79
Kudomi, K., 71
Kudryavtsev, V. A., 102
Kudryavtsev, V., 10, 102
Kulepov, V. F., 7, 76, 95
Kulsrud, R. M., 40
Kulsrud, R., 23, 40
Kulyk, Christine, 46
Kume, K., 71
Kupi, G., 30
Kursunoglu, Behram N., 17, 102, 104
Kusenko, A., 75
Kuzmichev, L. A., 7, 76, 95
Kuznetzov, V. E., 7, 76
Kwong, W., 82
Kyriazopoulou, S., 43, 75

L

Lachartre, D., 10
Lachaud, C., 93
Lafoux, H., 10
Lagomarsino, V., 79
Lai, C. H., 46, 47
Lai, D., 12, 56, 68, 75
Lamanna, E., 43, 76
LaMarche, P., 79
Lamare, P., 10
Lamb, D. J., 31
Laming, J. Martin, 49
Lamoureux, J. I., 81
Lampton, Michael, 25

Lande, Kenneth, 40
Lane, C., 43, 76
Languillat, J. C., 10
Lara, J. F., 49
Lattimer, J. M., 51
Laubenstein, M., 79
Laudinskaite, J. J., 76
Laugier, D., 10
Laugier, J. P., 10
Lawrence, D., 101
Lazarides, G., 51
Le Guen, Y., 10
Le Provost, H., 10
Le Van Suu, A., 10
Learned, J. G., 74, 95, 104
Lee, C. K., 40
Lee, H. K., 71
Lee, K., 74
Lee, S., 27
Lehtinen, Nikolai G., 83
Leich, H., 64, 81
Leiderer, P., 90
Lemoine, L., 10
Lesgourgues, Julien, 17
Letessier-Selvon, A., 93
Leuthold, M., 39, 64, 81
Levenfish, K. P., 47
Levin, D. S., 43, 76
Li, Jin, 80
Li, Xue-Qian, 54
Liebendoerfer, M., 22, 41
Liebendörfer, M., 31, 72
Liewer, K. M., 62
Lightfoot, P. K., 102
Lindahl, P., 39, 64, 81
Lindebaum, R. J., 69
Linke, F., 88
Liotard, P. L., 10
Lipari, P., 10, 43, 69, 76, 97-99
Lisi, E., 102, 107
Lisi, F., 53
Lissia, M., 37
Litchfield, P. J., 10
Little, L. T., 95
Liu, Q. Y., 42, 106
Liu, R., 76
Liubarsky, I., 39, 64, 81, 102
Ljaudenskaite, J. J., 7
Loaiza, P., 64, 81
Lobanov, A. E., 24
Lobashev, V. M., 21, 82
Loeb, Abraham, 93
Loeser, F., 79

Lola, S., 94
Lombardi, P., 79
Longley, N. P., 43, 76
Longo, M. J., 43, 76
Loparco, F., 43
Lopez, Robert E., 71, 72
Lopez, Robert Edwards Ph.D.
Loucatos, S., 10
Louis, W. C., 38
Lovzov, S. V., 7, 76, 95
Lowder, D. M., 39, 64, 81, 94
LSND Collaboration, 28, 38, 82
Lu, T., 42
Lubsadorzhiev, B. K., 95
Lubsandorzhiev, B. K., 7, 76
Ludlam, G., 76
Ludvig, J., 64
Lukash, V. N., 5
Luo, Q., 30
Lusignoli, M., 69

M

Ma, E., 39, 51
Maaroufi, F., 43, 76
Maartens, Roy, 11
Macelin, M., 10
MacRobert, Alan M., 87
Madsen, J., 64, 81
Magni, S., 79
Magnier, P., 10
Majorovits, B., 21, 31
Malvezzi, S., 79
Malyshkin, L., 23, 40
Mancarella, G., 24, 43, 76
Mandolesi, N., 65
Mandrioli, G., 43, 76, 81
Maniera, J., 79, 83
Mann, Alfred K., 74
Mannheim, K., 35, 52, 61, 66, 104
Manno, I., 79
Manuzio, G., 79
Manzoor, S., 43, 76
Marciniewski, P., 39, 64, 81
Margesin, Benno, 1
Margiotta Neri, A., 76
Margiotta, A., 43
Marini, A., 43, 76
Maris, Michele, 10
Maris, O., 81
Marshak, M. L., 62
Marshall, R., 102
Martello, D., 24, 43, 76

Martin, L., 10
Martino, J. Rodríguez, 64
Marx, G., 30
Marzari-Chiesa, A., 43, 76
Maschuw, R., 52, 53, 55
Masetti, F., 79
Massol, A., 10
Mathews, G. J., 30, 95, 102
Mathews, G., 62, 95
Matis, H. S., 64, 81
Matsuoka, K., 71
Mazéas, F., 10
Mazeau, B., 10
Mazure, A., 10
Mazziotta, M. N., 43, 76
Mazzotta, C., 76
Mazzucato, U., 79
McBride, P. L., 27
McDonald, A. B., 103
McDonald, Douglas Steven Ph.D., 91
McLaughlin, G., 74
McMillan, J., 10, 102
Mele, B., 79, 106
Melia, Fulvio, 68
Mendonça, J. T., 49, 69
Meroni, E., 79
Messer, O. E., 1, 22, 31, 41, 57, 72, 89
Mészáros, P., 18, 41, 108, 109
Meyer, B. S., 73, 74
Mezzacappa, A., 1, 6, 22, 31, 41, 57, 74, 83, 93, 99
Mezzacappa, Anthony, 72
Michael, D. G., 43, 76
Michael, D., 43, 76, 101
Migliozzi, P., 26
Mihalyi, A., 64
Mikheyev, S., 43, 76
Mikolajski, T., 39, 64
Milenin, M. B., 7, 76, 95
Miller, L., 43, 76, 101
Miller, T., 39, 64, 81
Millot, C., 10
Milyukov, V. K., 68
Minaeva, Y., 64, 81
Minakata, H., 6, 20, 40, 62, 109
Mintz, S. L., 17, 36, 55, 102
Miocinovic, P., 39, 64, 81
Miranda, O. G., 47, 57
Mirgazov, R. R., 7, 76, 95
Miyawaki, H., 71
Mock, P., 39, 64, 81
Mohapatra, R. N., 39, 55, 90, 100, 104
Mohri, M., 31

Mols, P., 10
Monacelli, P., 43, 76
Montalbán, J., 24
Montanet, F., 10, 33
Montanino, D., 94, 102, 107
Montanino, O., 53
Montaruli, T., 10, 42, 43, 45, 76, 80
Monteno, M., 43, 76
Moorhead, M. E., 33
Morales, Carmen, 25
Morel, J. P., 10
Moretti, M., 5
Morfín, J. G., 98
Moroz, A. V., 95
Morse, R., 39, 44, 64, 81
Moscoso, L., 9, 10, 44, 62
Moseiko, N. I., 7, 76, 95
Mosquera Cuesta, Herman J., 11
Mourão, A. M., 63
Mucke, A., 109
Mufson, S., 43, 76
Müller, E., 88
Murphy, A. St. J., 74
Murphy, B. T., 95
Musico, P., 79
Musser, J., 43, 76
Myers, J., 54

N

Nadyozhin, D. K., 51, 58, 100
Nagashima, Yorikiyo, 50
Nagataki, Shigehiro, 57, 102
Nakahata, M., 86
Nardi, Enrico, 74
Naudet, C. J., 62
Naumov, Vadim A., 55
Navas, S., 10
Neder, H., 79
Neff, M., 79
Nemtchenok, I. B., 25
Netikov, V. A., 7, 76
Neubauer, Mark Stephen Ph.D., 25
Neunhöffer, T., 64, 81
Newcomer, F. M., 64
Nicolò, D., 43, 76, 101
Niessen, P., 39, 64, 81
Nieto, M. M., 74
Nikiforov, S. A., 95
Nikonova, M., 38
Nisi, S., 79
Nolty, R., 43, 76
Normile, Dennis, 37

Novgorodov, A. F., 40
Novikov, V. M., 101
Nucciotti, A., 1, 54, 90
Nucita, A. A., 9
Nunokawa, H., 7, 20, 40, 92, 105
Nygren, D., 21, 64, 81

O

Oberauer, L., 13, 79
Obolensky, M., 79
Obridko, V. N., 82
O'Brien, M. S., 75, 98, 101
Oehler, C., 52
Ögelman, H., 64, 81
Ohsumi, H., 71
Okada, C., 43, 76
Olive, K. A., 72
Olivetto, C., 10
Orito, M., 46, 62, 95
Orlando, D., 9
Orth, C., 43, 76
Osipova, E. A., 7, 76, 95
Osteria, G., 43, 76
O'Sullivan, J. D., 4
Otsuki, Kaori, 31, 35, 102
Otten, E. W., 90, 100
Ouchrif, M., 43
Overduin, J. M., 2, 19

P

Pagerka, S., 17
Pakvasa, S., 52, 59
Pal, Palash B., 39
Palamara, O., 43, 76
Palanque-Delabrouille, N., 10
Palladino, V., 26, 29
Pallares, A., 10
Pallavicini, M., 80
Panella, O., 84
Panfilov, A. I., 7, 76, 95
Panov, I. V., 100
Pantis, G., 105
Papadopoulos, P., 88
Papp, L., 80
Para, A., 40
Parente, G., 65
Parfenov, Yu. V., 7, 76, 95
Parlati, S., 76
Päs, H., 2
Pasquali, L., 10, 82
Pastor, S., 17, 53, 82

Patera, V., 43, 76
Paterno, L., 99, 103
Pati, Jogesh C., 109
Patrizii, L., 43, 76, 81
Pavan, Maura, 1
Pavlov, A. A., 7, 76
Payre, P., 10
Pazzi, R., 43, 45, 76
Peccei, R. D., 54
Peck, C. W., 43, 76
Perasso, L., 80
Pereira, J. G., 87
Peres, O., 7
Pérez de los Heros, C., 39, 64, 81
Pérez-Mercader, Juan, 25
Perlmutter, Arnold, 17, 102
Perrin, P., 10
Perrone, L., 43, 55
Pessina, Gianluigi, 1
Petcov, S., 51, 101
Pethick, C. J., 58
Petrera, S., 43, 76
Pfaudler, J., 70
Picchi, P., 20
Piepke, A., 20, 101
Pietropaolo, F., 20, 96
Pigmatel, Giorgio, 1
Pilaftsis, A., 11
Pinsonneault, M. H., 35, 85
Pinto, P., 3, 5
Pisanti, O., 37
Pistilli, P., 43, 76
Pittalwala, S., 101
Pizzochero, Pierre M., 62
Plaian, A., 81
Plischke, P., 52
Pliskovsky, E. N., 7, 76
Pocar, A., 80
Pohl, A., 10
Pohl, M., 59
Poinsignon, J., 10
Pokhil, P. G., 7, 76
Polesello, G., 78
Pons, J., 51
Popa, L., 65
Popa, V., 43, 76, 81
Popova, E. G., 7, 76
Porrata, R., 39, 64, 81
Pospelov, M., 72
Postma, M., 1
Potekhin, A. Y., 58
Potheau, R., 11
Potter, D., 39

Pourkaviani, M., 55
Prakash, M., 51
Previtali, Ezio, 1
Price, P. B., 6, 39, 64, 78, 81
Protheroe, R., 9, 30, 66
Protheroe, Ray, 60
Provost, J., 85
Pruet, Jason, 65
Przybylski, G., 39
Przyrembel, M., 90
Pugliese, G., 100
Pulido, J., 77, 84

Q

Qian, Y.-Z., 48, 56, 57, 68
Queinec, Y., 11

R

Racca, C., 11, 106
Rachen, J. P., 18, 66, 108, 109
Raffelt, G. G., 41, 45, 92
Raghavan, R., 80
Raimondo, G., 90
Rainó, A., 43, 76
Rajpoot, Subhash, 52
Ralston, John P., 42
Rampp, Markus, 89
Ranucci, G., 80
Rapp, J., 52
Rau, W., 80
Rawlins, K., 64, 81
Raymond, M., 11
Razeto, A., 80
Reddy, S., 51, 61
Reed, C., 64, 81
Rees, M. J., 41
Reichenbacher, J., 52
Reno, M. H., 10, 82
Resconi, E., 80
Reya, E., 22
Reynoldson, J., 43, 76
Rhode, W., 43, 64, 81, 90
Ricci, B., 3, 85, 87
Ricciardi, S., 52
Richards, A., 39, 64
Richter, S., 39, 64, 81
Riedel, T., 80
Riordan, Michael, 69
Ritchie, B., 101
Rivera-Rebolledo, J. M., 11
Rivin, Yu. R., 82

Robertson, R. G. H., 3, 21, 42
Robinson, L. J., 67, 93
Roca-Cortes, T., 85
Rodríguez Martino, J., 81
Rolin, J. F., 11
Romanino, A., 28
Romenesko, P., 39, 64, 81
Ronga, F., 43, 45, 76, 80
Roos, M., 104
Rosa, L., 37
Rosen, S. P., 8, 82
Ross, D., 64, 81
Rostovtsev, A., 100
Roth, J., 67
Roulet, Esteban, 61
Roxburgh, I. W., 32, 85
Rozanov, M. I., 7, 76
Rubinstein, H., 39, 64
Rubizzo, U., 76
Rubzov, V. Y., 7, 76
Rudenko, V. N., 68
Ruf, C., 52
Ruffert, M., 13
Ruffini, R., 99
Rysavý, M., 40

S

Sabelnikov, A., 80
Sacquin, Y., 11
Saggese, P., 80
Sahu, S., 93
Salis, A., 106
Salmonson, Jay D., 30, 31, 43
Saltzberg, D. P., 62
Salvo, C., 79, 80
Sanchez, L. A., 99
Sánchez, N. M., 46
Sander, H.-G., 64, 81
Sanford, Robert Ellis, Jr. Ph.D., 88
Santacesaria, R., 29
Sanzgiri, A., 76
Sarcevic, I., 10, 82
Sarkar, S., 17
Sarkar, U., 24
Sassaroli, E., 48
Sato, K., 20, 23, 29, 45, 71
Satriano, C., 43, 76
Satta, L., 43, 76
Scapparone, E., 43, 76
Scardaoni, R., 79, 80
Scargle, J. D., 34, 79, 107
Scarsi, L., 98

Schatzman, E., 24
Scheider, T., 64
Schlattl, H. Ph.D., 104
Schlickeiser, R., 59
Schmidt, T., 39, 64, 81
Schnagl, J., 19
Schneider, D., 64, 81
Schneider, E., 39, 64
Schnürer, F., 52
Schoenert, S., 79, 80
Scholberg, K., 44, 76
Schuhbeck, K., 79, 80
Schuller, J.-P., 11
Schuster, C., 59
Schuster, W., 11
Schwarz, R., 39, 64, 81
Schwetz, T., 28
Sciama, D. W., 17, 20, 31, 67, 90
Scioscia, G., 53
Sciubba, A., 44, 76
Scott, W. G., 36, 90
Seahra, S. S., 2, 19
Seckel, D., 36
Sedrakian, A., 58
Segre, G., 75
Seidel, H., 79, 80
Seligmann, B., 52
Semikoz, D. V., 37, 53, 63
Semikoz, V., 47, 57, 82
Serra, P., 44
Serra-Lugaresi, P., 76
Sethi, Shiv K., 20
Severi, M., 76
Shaevitz, M. H., 24, 82
Shapiro, Stuart L., 75
Shi, X., 59, 64
Shibahashi, H., 7, 37, 83-85
Shimada, Yoriaki, 109
Shimizu, Tetsuya M., 23
Shiomi, S., 71
Shirai, J., 37
Shukla, P. K., 49
Shutt, T., 79, 80
Sigismondi, C., 99
Sigl, G., 27, 71
Silk, Joseph, 17
Silva, L. O., 49, 69, 70
Silvestri, A., 64, 81
Simgen, H., 79, 80
Simkovic, F., 105
Sinclair, D., 27
Singh, S. K., 63
Sioli, M., 44, 76

Sisti, Monica, 1
Sitta, M., 44, 76
Skorokhvatov, M., 79, 80
Sky Publishing Corp. Team, 67
Smirnov, A. Yu., 42, 45
Smirnov, O., 79, 80
Smit, J. M., 12
Smith, D., 26
Smith, N. J. T., 102
Smith, P. F., 28, 29, 74, 102
Smolnikov, A. A., 25
SNEWS Collaboration, 84
SNO Collaboration, 103
Sokalski, I. A., 7, 76
Solarz, M., 39, 64, 81
Soler, Tomás, 6
Song, Jeonghyeon Ph.D., 87
Sonnenschein, A., 79, 80
Sotnikov, A., 79, 80
Sotnikova, R., 38
Spada, M., 49, 66
Spalek, A., 40
Spiczak, G. M., 39, 64, 81
Spiering, C., 7, 39, 64, 81
Spiering, Ch., 76
Spinelli, P., 44, 76
Spinetti, M., 44, 76
Spooner, N., 11, 102
Spurio, M., 38, 44, 76
Stancu, I., 25, 28
Stanev, Todor, 7, 10, 105
Starinsky, N., 64, 81
Stecker, F. W., 108
Steele, D., 64, 81
Stefanov, L., 81
Steffen, P., 64, 81
Steidl, M., 52
Steinberg, R., 44, 76
Stodolsky, L., 91
Stojkovic, Dejan Ph.D., 50
Stokstad, R. G., 64, 74, 81
Stolarczyk, T., 11
Stone, J., 44, 76, 78
Strayer, M. R., 6, 99
Strecker, H., 21
Streicher, O., 7, 39, 64, 76, 81
Stucken, I., 52
Studenikin, A., 24, 68
Sturrock, P. A., 8, 14, 15, 34, 78, 79, 87, 107
Sudhoff, P., 81
Sukhotin, S., 79, 80
Sulak, L. R., 44, 76
Sullivan, G., 86

Sumiyoshi, Kohsuke, 35
Sumner, T. J., 102
Sun, Q., 39, 64
Sundaresan, M. K., 13
Super-Kamiokande Collaboration, 86
Surdo, A., 24, 44, 76
Surman, Rebecca, 28, 62
Suslin, V. M., 83
Suzuki, A., 87
Suzuki, H., 16, 35, 56
Suzuki, Y., 54, 84, 86, 100
Svet, V. D., 74
Svoboda, R., 61
Swift, T. P., 73

T

Tabary, A., 11
Taboada, I., 64, 81
Tagoshi, Hideyuki, 31
Tajima, T., 46
Takahashi, Y., 31
Takahisa, K., 71
Takata, M., 7, 83, 84
Talby, M., 11
Tanimoto, M., 31, 37, 94
Tao, C., 11
Tarashansky, B. A., 7, 76
Tarlé, G., 44, 76
Tartaglia, R., 79, 80
Tayalati, Y., 11, 67
Teller, Edward, 66
Terasawa, Mariko, 35
Testera, G., 79, 80
Teves, W. J. C., 91
AMANDA Collaboration, The, 64, 94
ANTARES Collaboration, The, 33
HIRES Collaboration, The, 80
K2k Collaboration, The, 100
LSND Collaboration, The, 25
MACRO Collaboration, The, 10, 34, 38, 42, 45
MACRO collaboration, The, 80
NOMAD Collaboration, The, 78
SOUDAN 2 Collaboration, The, 62
Super-KAMIOKANDE Collaboration, The, 10,
    25, 61, 84, 86
Thielemann, F.-K., 22, 72
Thollander, L., 11, 39, 64, 81
Thompson, Todd A., 3
Thon, T., 7, 39, 64, 76, 81
Thorsson, V., 58
Tietze, G., 90
Tilav, S., 39, 64, 81

Timmermans, R. G. E., 58
Togo, V., 44, 76
Tokuhisa, A., 46
Tomàs, R., 105
Torrente-Lujan, E., 45
Torres, M., 50
Totani, T., 24, 29
Tothi, G., 7
Totsuka, Y., 25, 54
Tracy, D., 101
Trapero, Joaquín, 25
Triay, R., 11
Triginer, Josep, 11
Tsiklauri, D., 32
Tsiklauri, David, 32, 66
Tsintsadze, L. N., 12, 104
Tsintsadze, N. L., 17
Tsou, Sheung Tsun, 70
Tsun, T. S., 70
Tsytovich, V. N., 13
Tupper, G. B., 69
Turck-Chièze, S., 70, 85, 103
Tutukov, A. V., 36
Tytell, David, 87
Tzvetanov, T., 11

U

Uehara, Y., 73
Ullman, Jack, 40
Ulrich, R. K., 85
Umar, A. S., 6, 99
Usechak, N., 64
Ustyugov, S. D., 83

V

Vagins, M., 74
Vaitaitis, Arturas Genrikas Ph.D., 80
Vakili, M., 44
Valdy, P., 11
Valeanu, V., 81
Valente, V., 76
Valle, J. W. F., 7, 47, 54, 57, 76, 82, 103, 105
van Dalen, E. N. E., 58
van den Horn, L. J., 12
Vander Donckt, M., 46, 64, 81
Vannucci, F., 59
Varieschi, Gabriele Umberto Ph.D., 69
Vasiliev, S. I., 25
Vasiljev, R., 7, 76
Vauclair, Sylvie, 32
Vázquez, R. A., 101

Vergados, J. D., 94
Vernin, P., 11
Vernon, W., 74
Vietri, M., 65, 106
Vigeolas, E., 11
Vignaud, D., 11, 27, 54
Vilanova, D., 11
Vilela, E., 44
Villante, F. L., 5
Vinogradov, M. P., 68
Viollier, R. D., 32, 66, 69, 104, 109
Vissani, F., 104
Vital, A., 101
Vitale, S., 1, 79, 80
Vitiello, G., 32
Vogel, P., 12, 20, 38, 39, 48, 101
Vogelaar, R., 79, 80
Volkas, R. R., 14, 39, 68
Volkov, D. V., 70
von Feilitzsch, F., 79, 86
von Hentig, R., 79
Voskresensky, Dmitri N., 45
Vuilleumier, J.-L., 20

W

Wagner, D. J., 36
Walck, C., 39, 64, 81
Walter, C. W., 44, 76
Walther, G., 8, 67, 78, 79, 107
Wanajo, Shinya, 31, 102
Wang, Y. F., 20, 101
Wang, Y., 20, 95, 100, 101
Wark, D., 11
Wasserburg, G. J., 48
Watanabe, Satoru, 37, 85
Watson, A. A., 101
Waxman, E., 7, 30, 33, 42, 49, 66, 93
Webb, R., 44, 76
Weber, M. A., 14, 15, 107
Weiler, T., 17, 18, 25, 27, 36, 70, 77
Weinheimer, C., 50, 64, 81, 90, 100
Wesson, P. S., 2, 19
Wheatland, M. S., 8, 78, 79, 107
Whisnant, K., 51
Whisnant, Kerry, 8, 105
White, D. H., 27
Widom, A., 105
Wiebusch, C. H., 64, 81
Wiebusch, C., 39, 64, 81
Wilczek, Frank, 46
Wildenhain, Paul S., 40
Wilfred, R., 101

Wilkerson, J. F., 21
Wilkes, R. Jeffrey, 78
Williams, D., 62
Wilson, J. R., 29-31, 43
Wilson, Robert M., 17
Wilson, T. L., 55
Wischnewski, R., 7, 39, 64, 76, 81
Wissing, H., 64, 81
Wittich, Peter Ph.D., 27
Wodecki, A., 17
Wojcik, M., 79, 80
Wolf, J., 52, 101
Wong, H. T., 39, 80
Wong, Y. Y. Y., 14, 39
Woodahl, Brian Arvid Ph.D., 107
Woosley, S., 54
Woschnagg, K., 39, 64, 78, 81
Wu, W., 64, 81

X

Xing, Zhi-Zhong, 42

Y

Yakovlev, D. G., 47, 48, 58
Yamada, S., 16, 23, 35, 40, 56

Yanagida, T., 39
Yanagisawa, C., 76
Yang, S., 101
Yashin, I. V., 7, 76
Yasuda, O., 62, 91, 104
Yodh, G., 39, 64, 81
Yoshida, S., 27, 71, 74
Young, S., 4, 64, 81, 101
Young, T., 3, 5, 59, 93

Z

Zach, J. J., 74
Zaimidoroga, O., 79, 80
Zakharov, Y., 79, 80
Zanotti, Luigi, 1
Zas, E., 23, 65, 74, 99, 101
Zeitnitz, B., 52
Zeldovich, O., 20
Zen, Mario, 1
Zghiche, A., 11
Zhang, C. M., 87
Zheleznykh, I. M., 74
Zuber, K., 37, 90
Zukanovich Funchal, R., 91
Zuluaga, Jorge I., 74
Zúñiga, J., 11

Subject Index

A

Active Galactic Nuclei (AGN), 4, 7, 9, 33, 36, 60, 108

active-sterile neutrino oscillations, 14

algorithm, 5, 81, 83

alpha magnetic spectrometer (AMS), 98

amplitude, 8, 19, 24, 28

anisotropic, 23, 47, 57, 75

Antarctic Muon and Neutrino Detector Array (AMANDA), 4, 6, 35, 39, 55, 63, 64, 78, 81, 90, 94, 95, 108

antineutrinos, 2, 9, 13, 16, 18, 29, 38, 47, 49, 55, 66, 75, 89, 104, 106, 107

astronomical observations, 12, 51

asymmetric neutrino flux, 12

automated program generator, 10

B

baryogenesis, 45

baryons, 13, 51, 52, 58, 88

big bang nucleosynthesis (BBN), 36, 45, 62

Boltzmann solver, 3, 56, 89

C

central engines, 6

charged-current (CC), 25, 27, 38, 81, 92

chemical mixing, 24

correlation analysis, 15

Cosmic Microwave Background (CMB), 33, 39, 44, 52, 62, 65, 71, 72, 96, 98

Cosmic Microwave Background Radiation (CMBR), 44

Cosmic Neutrino Astronomy, 33

cosmic neutrino, 23, 90

cosmic-rays, 33

cosmological models, 5, 44, 96

cosmological neutrino, 19

D

decoupling layer, 12

decoupling, 12, 87, 96

deleptonization, 7, 61, 69, 99

density profile, 7, 38, 42, 86

Dirac neutrino, 55

E

electron neutrino, vii, 2, 7, 15, 17, 25, 27, 93, 104, 105

electrons, vii, 27, 38, 47, 49, 52, 58, 59, 67, 69, 77, 88, 94, 95, 97, 100, 102

emergent spectra, 4, 5, 56

energy deposition, 13, 26, 43, 53, 57, 65, 77, 88

energy source, 2

evolution, 5, 7, 13, 14, 16, 26, 29, 30, 33, 46, 51, 60-62, 66, 67, 72, 75, 77, 85, 98-101, 103

F

fermi gases, 16

fireball, 13, 30, 41, 48, 66

flavor oscillations, 3, 92

fluctuation-dissipation, 47

G

galactic center (GC), 68

galaxy clusters, 5

gamma-ray burst(s) (GRB(s)), 4, 7, 13, 30, 32, 33, 41-45, 47, 48, 49, 52-54, 65, 66, 77, 80, 100, 108

gravitational redshifts, 3

gravitational wave, 11, 13

H

hard neutrino spectra, 4
heating mechanism, 1, 16, 46
helioseismic constraints, 3, 37
helioseismic measurements, 2, 18
Helioseismology, 15, 32, 84, 87
high energies, 4, 30, 36, 60
high energy (HE), 3, 4, 7, 9, 27, 33, 35, 36, 42, 43, 45, 49, 60, 63, 64, 65, 66, 71, 72, 78, 80, 83, 91, 95, 101, 104, 107
high-energy neutrinos, 4, 7, 9, 33, 48, 55, 60, 64, 66
high-sensitivity spectrometer, 25
hot dark matter (HDM), 6, 24, 27, 79, 105, 106
hot particles, 5
hydrodynamical flow, 13
hydrostatic equilibrium, 16

I

inelastic collisions, 73, 100
infra-red light (IR), 36, 44
interstellar medium (ISM), 42, 46, 60, 65, 68, 106

L

luminosities, 2, 5, 13, 22, 57, 61, 65, 72, 88
Luminosity Function (LF), 100
luminosity, 2, 9, 13, 17, 23, 29, 33, 35, 47, 49, 57, 59, 65, 85, 86, 88, 89, 93, 98, 108

M

magnetic field, 8, 15, 20, 35, 49, 55, 56, 57, 58, 69, 75, 77, 78, 79, 82, 108
mass determination, 1, 38
massive neutrino, 5, 28, 106
Mikheyev-Smirnov-Wolfenstein (MSW), 7, 8, 14, 20, 40, 53, 78, 82, 102
modulation, 14, 15, 34, 78, 79, 82, 102
Monopole Astrophysics and Cosmic Ray Observatory (MACRO), 3, 10, 34, 38, 42-45, 54, 76, 80
multigroup Boltzmann (MGBT), 1
multigroup flux-limited diffusion (MGFLD), 1, 7, 56, 89, 99
muon neutrino flux, 3, 18
muon neutrinos, 4, 27, 55, 60, 63, 82, 83, 104
muons, 3, 6, 43-45, 63, 65, 69, 72, 77, 80, 82, 83, 94-97

N

neutrino detectors, 2, 7, 12, 59, 72
neutrino emission, 13, 23, 35, 45, 47, 56, 58, 59-62, 64, 68, 69, 75, 88, 95, 98, 101
neutrino luminosity, 23, 35
neutrino masses, 24, 31, 36, 38, 42, 50, 59, 61, 93, 104
neutrino telescope, 11, 35, 39, 42, 55, 76, 81, 95
neutrino transport, 1, 3, 5, 7, 22, 26, 41, 56, 57, 72, 89, 99
neutrino-nucleon scattering, 4, 69
neutrinosphere geometry, 11
nuclei, 6, 9, 16, 23, 28, 29, 31, 35, 36, 48, 57, 58, 59, 60, 66, 73, 78, 89, 95, 98, 100, 101, 102, 106
nucleons, 16, 18, 41, 47, 61, 63, 92
nucleosynthesis, 28, 31, 35, 47, 51, 57, 62, 73, 78, 96, 102, 105

O

Observatory for Multiflavor Neutrino Interactions from Supernovae (OMNIS), 21, 28, 74, 93
one mass scale dominance (OMSD), 5
optical modules (OMs), 39, 81
oscillation hypothesis, 7

P

particle physics, 6, 21, 32-34, 44, 46, 54, 55, 62, 71, 75, 78, 80, 91
phase transition, 47
photomultiplier tubes, 11, 55, 94
photons, 18, 20, 26, 44, 49, 60, 66, 68, 69, 81, 84, 91, 93, 95, 106
positrons, 47, 81
postbounce, 1, 4, 6, 22, 72
pre-white dwarfs (PWDs), 75, 98
proto-neutron star (PNS), 4, 11, 23, 35, 41, 61, 68, 99, 100, 102
protoneutron, 4, 22, 51, 61

R

radiative zone, 15, 78, 79, 108
radiatively decaying, 20, 25
rapid rotation, 11
real time measurements, 1
recoil energy, 21
resonance statistic, 14, 15
resonant spin-flavor, 8, 20
rotation rate, 15, 78, 108

S

scattering, 12, 13, 18, 21, 24, 25, 38, 41, 47, 58, 66, 68, 80, 88, 92, 94, 97, 100, 101, 102, 106
shock revival, 16
significant fluxes, 41, 98
solar center, 24
solar core, 2, 3, 71, 103, 108
solar cycle, 17
solar electron, 12, 17, 91, 104
solar models, 3, 8, 16, 37, 70, 83, 85, 87, 91
solar neutrino fluxes, 2
solar neutrino oscillations, 8, 10, 25, 85, 105
solar neutrino problem, 7, 12, 20, 24, 25, 27, 32, 38, 67, 77, 80, 81, 82, 85, 86, 102, 105, 107
solar neutrino units (SNU), 17, 34, 40
solar neutrinos, 3, 17, 18, 25, 27, 28, 33, 40, 80, 86, 87, 88, 91, 97, 104, 105
solar radiative zone, 8
Southeastern Association for Research in Astronomy (SARA), 73
spherical models, 23
spontaneous symmetry breaking (SSB), 47
Standard Solar Model (SSM), 76, 105
Star Formation Rates (SFR), 100
static approximation, 14
static neutrino atmospheres, 5
stellar matter, 13, 100
sterile neutrinos, 8, 11, 17, 39, 52, 53
Sudbury Neutrino Observatory (SNO), 11, 25, 27, 38, 84, 91, 92, 97, 103
Sun's internal rotation, 14, 15
sunspot, 17, 22, 67
Super-Kamiokande (SK), 5, 8, 10, 19, 21, 25, 29, 37, 38, 50, 52, 76, 78, 84, 86-88, 92, 93, 97, 100

supernova, 1, 11, 12, 13, 16, 21-26, 28-31, 33, 38, 41, 45, 46, 48, 49, 56, 57, 59, 60, 64, 67, 69, 72-75, 78, 81, 84, 89, 92, 93, 102, 104, 105

T

tau neutrinos, 4, 21, 29, 92, 93, 104
thermal detectors, 1
total mass, 1, 16, 21
transport methods, 2

U

Ultra-high-energy (UHE), 22, 23, 64, 67, 74, 79, 106, 108
United States, 4

V

variable planetary nebula nuclei (PNNV), 98
velocity, 4, 12, 16, 56, 57, 65, 75, 78, 86, 89, 102
vertical axis, 6, 97
viscous fluid, 88

W

weak magnetism, 4
weakly interacting massive particle (WIMP), 21, 31, 55, 82
wrong-helicity, 9

X

X-ray, 14, 15, 22, 60, 106